OpenShift Multi-Cluster Management Handbook

Go from architecture to pipelines using GitOps

Giovanni Fontana

Rafael Pecora

BIRMINGHAM—MUMBAI

OpenShift Multi-Cluster Management Handbook

Group Product Manager: Rahul Nair

Publishing Product Manager: Meeta Rajani

Senior Content Development Editor: Sayali Pingale

Technical Editor: Shruthi Shetty

Copy Editor: Safis Editing

Book Project Manager: Neil Dmello

Proofreader: Safis Editing

Indexer: Rekha Nair

Production Designer: Alishon Mendonca

Marketing Coordinator: Nimisha Dua

First published: October 2022

Production reference:1141022

Published by Packt Publishing Ltd.

Livery Place

35 Livery Street

Birmingham

B3 2PB, UK.

978-1-80323-528-8

www.packt.com

To my greatest treasures here on earth: my sweet and smart son, Pedro, and my lovely wife, Jeila, who tolerated my "good humor" after long nights awake writing this book! :)

You both gave me amazing support and motivation all the time! I wouldn't be able to accomplish this without you! Both of you are the inspiration I have to do my best every day! Love you so much!

– Giovanni Fontana

Some people leave and let a little of themselves, others come and take a little of us, but only those who are substantially part of our lives give us the strength to continue, so I dedicate this book affectionately to my smart son, Davi, to whom I dedicate unconditional love, and to my love and inspiration, my wife, Fabiana.

– Rafael Pecora

Foreword

Installing, configuring, and using OpenShift can require you to take several complex steps. Understanding all the necessary steps and acquiring knowledge so that your environment is perfectly installed can take time. With the help of our authors *Giovanni (Gigio)* Fontana and *Rafael Pecora* who know the subject, and have years of experience in everything from architecture design to implementation and support, this path can be shortened.

This book contains not only the necessary steps to install, configure, and use OpenShift, but also information and helpful tips, so that everything is perfectly installed.

Giovanni is always loved by everyone around him, and as a perfectionist who enjoys applying himself to complicated tasks; he likes things well done.

Pecora is a prodigy, a true smart guy, and at the same time he is self-taught; he learns things with a level of ease that is unique to him.

Marcos Amorim

Sr. Support Engineer

Red Hat

Contributors

About the authors

Giovanni Fontana is a solution architect working for large companies in the US Northeast region from different industries such as finance and the health sector. He is a Red Hat Certified Architect, owning many certificates covering RHEL, OpenShift, Ansible, and others. In the cloud and DevOps domain, he has been working with activities such as process assessment and design; pre-sales, architecture design, and implementation of container-based platforms such as Red Hat OpenShift Container Platform, hybrid/multi-cloud management tools; automation design and implementation using Ansible, and others. Before his current position, he was a principal consultant with much hands-on experience, providing services for many large companies and also acting as a technical leader for a talented team of consultants at Red Hat in Brazil. Now as an OpenShift specialist solution architect, he helps customers to solve pain points and reach their targets by adopting Red Hat's open source technologies.

At Red Hat, Giovanni has been recognized with important awards such as Red Hat President's Club, Top Gun, and Champions awards.

Rafael Pecora started his first journey with technology at the age of 15 and since then, he has been dedicating himself to technology. He has more than 23 years dedicated to IT, a profession that is always evolving and shaping itself on cutting-edge innovation. Passionate about IT, he has focused his entire career on IT infrastructure, from traditional data centers to current cloud technologies. Currently, he develops his professional activity as a cloud solution architect in container technology and public clouds, with the aim of bringing his knowledge and improvement to clients that work in the financial, health and insurance, mid-market, and public sector, and helping them adopt hybrid cloud infrastructure with better resiliency for their customers.

At Red Hat, Rafael has achieved more than 40 projects as an OpenShift consultant, acting on many industry projects, and has been recognized by customers and also recognized with Service Star and Champions awards.

About the reviewers

Rafael Aracely Delfino Sales works at Red Hat as an engineer across Latin America. He is a specialist in the Red Hat OpenShift Container Platform and has been helping clients modernize their infrastructure since 2018.

I am grateful to Giovanni Fontana and Rafael Pecora, who invited me to review this book and trusted my technical skills. It was an amazing experience to be part of this project.

Andres Sacco is a technical leader at TravelX and he has experience with many different languages, such as Java, PHP, and Node.js. In his previous job, Andres helped find alternative ways to optimize data transfer between microservices, which reduced the infrastructure cost by 55%. Before introducing these optimizations, he investigated different testing methods for better coverage than the unit tests of the microservices. He also dictated internal courses about new technologies, and he has written articles on Medium. Andres is a co-author of the book *Beginning Scala 3*, from Apress.

Table of Contents

3

Multi-Tenant Considerations   49

4

OpenShift Personas and Skillsets   61

Part 2 – Leverage Enterprise Products with Red Hat OpenShift

5

6

7

OpenShift Network   147

8

OpenShift Security   171

Part 3 – Multi-Cluster CI/CD on OpenShift Using GitOps

9

OpenShift Pipelines – Tekton   189

10

OpenShift GitOps – Argo CD   231

11

OpenShift Multi-Cluster GitOps and Management   265

Part 4 – A Taste of Multi-Cluster Implementation and Security Compliance

12

OpenShift Multi-Cluster Security   313

13

OpenShift Plus – a Multi-Cluster Enterprise Ready Solution   359

14

Building a Cloud-Native Use Case on a Hybrid Cloud Environment   379

Part 5 – Continuous Learning

15

Preface

We are living in a world of such big and rapid changes. In the IT industry, things such as **DevOps**, **the cloud**, **digital transformation**, **containers**, and **Kubernetes** have emerged in the last few years and dramatically changed the way we develop, build, test, and deploy applications.

Among the big set of new technologies, only a few of them have become a sort of *consensus* in the industry, including the cloud, containers, and Kubernetes. Over the last few years, the authors of this book have worked with many customers that are on this cloud and container adoption journey – *almost all of them have adopted some sort of Kubernetes distribution and one or more cloud providers*. Most of them are *experiencing the benefits* of this adoption but also have to deal with *many new challenges*, such as *maintaining standard environments, keeping them secure, and keeping costs under control*.

That is why we decided to focus this book not *only* on OpenShift, which is one of the market leaders in enterprise Kubernetes, but *also* on **multi-cluster management**. This book is the result of *years of experience in the field*, designing architectures and deploying and operating OpenShift clusters, which we have tried to turn into words, pages, and chapters!

We are going to cover in this book answers to questions such as, *what are the different architectural choices I have with OpenShift? How do I design the right architecture for my case? What are the different personas related to OpenShift, their main responsibilities, and their challenges?* We will also walk through **OpenShift deployment, troubleshooting, network, and security**. CI/CD and **GitOps** are also covered in this book with some *practical examples* that you can use to learn how they can help you to increase the maturity level of your build and deployment process. Finally, we will go over some tools, such as **Advanced Cluster Management** and **Advanced Cluster Security**, that will help you to *manage and secure multiple OpenShift clusters from a central standpoint*.

Who this book is for

This book is intended to be helpful for any professional that is involved with Kubernetes and OpenShift. It is especially useful for software engineers, infrastructure operators, developers, site reliability engineers, and security engineers, but should also be relevant to enterprise architects and CXO decision makers. Are you interested in knowing more about OpenShift architecture, deployment, security, GitOps, pipelines, multi-cluster management, and security? *If so, this book was carefully made for you!*

What this book covers

Chapter 1, Hybrid Cloud Journey and Strategies, discusses the main challenges of public cloud adoption and explains what OpenShift is, and why it can help to deal with those challenges to achieve success in the business, culture, and technical aspects of hybrid cloud adoption.

Chapter 2, Architecture Overview and Definitions, walks through the main concepts related to Kubernetes and OpenShift architecture to help you decide on the best path you may take.

Chapter 3, Multi-Tenant Considerations, helps you to work with multi-tenancy on OpenShift to provide multiple environments to multiple teams, with proper isolation for every case.

Chapter 4, OpenShift Personas and Skillset, looks at changes that can be made in a company structure to help people adapt to their roles and responsibilities.

Chapter 5, OpenShift Deployment, is a complete hands-on guide to installing and using OpenShift.

Chapter 6, OpenShift Troubleshooting, Performance, and Best Practices, demonstrates some of the most common issues with OpenShift usage.

Chapter 7, OpenShift Network, explores OpenShift's network layers, such as the Open vSwitch, as well as the north-south and east-west traffic concepts, and the different types of TLS configurations for OpenShift routes.

Chapter 8, OpenShift Security, presents some of the most important concepts of security such as container security, authentication and authorization, identity providers, **Role-Based Access Control (RBAC)**, certificates, etcd encryption, container and network isolation, the Red Hat container catalog for image certification, and vulnerability protection.

Chapter 9, OpenShift Pipelines – Tekton, introduces OpenShift Pipelines, a Kubernetes native CI/CD pipeline tool based on Tekton. This chapter contains the main concepts, the installation process, and a hands-on lab to learn using a practical approach.

Chapter 10, OpenShift GitOps – ArgoCD, expands the deployment capabilities by adding GitOps and ArgoCD. This chapter covers GitOps concepts, OpenShift GitOps installation, and a hands-on lab.

Chapter 11, OpenShift Multi-Cluster GitOps and Management, deep dives into hybrid/multi-cloud concepts and the main concerns about adopting multiple clusters. This chapter describes what Red Hat Advanced Cluster Management is, its installation, and how to use it to manage multiple Kubernetes clusters from a central console.

Chapter 12, OpenShift Multi-Cluster Security, expands the concepts from *Chapter 8, OpenShift Security*, focusing on multi-cluster security concerns. This chapter covers the Red Hat Advanced Cluster Security tool features, such as risk management, vulnerabilities, violation, policies, compliance, and configuration management, along with installation, configuration, and usage instructions.

Chapter 13, OpenShift Plus – a Multi-Cluster Enterprise Ready Solution, introduces Red Hat Quay as an enterprise image registry option, and it also discusses the benefits of the OpenShift Plus offering as a great option for enterprises looking for consistency and portability in a hybrid/multi-cloud environment.

Chapter 14, Building a Cloud-Native Use Case on a Hybrid Cloud Environment, introduces a complete practical example. This chapter transposes all the concepts in the book using a step-by-step hands-on guide to show how to build and deploy an application using most of the features covered throughout the book: OpenShift Pipelines (Tekton), OpenShift GitOps (ArgoCD), Advanced Cluster Management, Quay, and Advanced Cluster Security.

Chapter 15, What's Next, offers suggestions for the next steps to take to keep learning and going even deeper into OpenShift through training and certifications.

To get the most out of this book

Throughout this book, you will run commands and use command-line tools. We recommend you use a Mac or Linux workstation, such as Fedora or Ubuntu – although some of the CLI tools also have versions for Windows, the command examples used in this book assume the usage of a Linux workstation, and Mac should also work fine.

For some sections of this book, an AWS account is necessary; however, it should not be difficult to adapt the examples for other cloud providers. You can sign up for a free trial on AWS here: `https://aws.amazon.com/`.

Command-lines used in the book	OS requirements
OpenShift Client	Windows, macOS, and Linux (Mac and/or Linux preferred)
OpenShift Installer	macOS and Linux
CodeReady Containers (CRC)	Windows, macOS, and Linux (Mac and/or Linux preferred)

Although we have done our best to make this book as didactic and inclusive as possible, you will probably get more out of it if you already have a basic knowledge of containerization and Kubernetes – no advanced knowledge is needed though.

Download the color images

We also provide a PDF file that has color images of the screenshots/diagrams used in this book. You can download it here: `https://packt.link/C8KLC`.

Download the example code files

You can download the example code files for this book from GitHub at `https://github.com/`
`PacktPublishing/OpenShift-Multi-Cluster-Management-Handbook`. If there's
an update to the code, it will be updated on the existing GitHub repository.

We also have other code bundles from our rich catalog of books and videos available at `https://`
`github.com/PacktPublishing/`. Check them out!

Conventions used

There are a number of text conventions used throughout this book.

`Code in text`: Indicates code words in text, database table names, folder names, filenames, file
extensions, pathnames, dummy URLs, user input, and Twitter handles. Here is an example: "For this
step, we are going to use the `quarkus-build` pipeline, which you will find in the `chapter14/`
`Build/Pipeline/quarkus-build.yaml` file."

A block of code is set as follows:

```
SELECT
     COUNT(DISTINCT Column1) AS Count_column1,
     System.TIMESTAMP() AS Time
FROM Input TIMESTAMP BY TIME
GROUP BY
     TumblingWindow(second, 2)
```

Code output or a command line entry is set as follows:

```
npm install applicationinsights
```

Bold: Indicates a new term, an important word, or words that you see onscreen. For example, words
in menus or dialog boxes appear in the text like this. Here is an example: "After you have created the
AWS credential, access the **Infrastructure | Clusters** feature and click on the **Create cluster** button."

> **Tips or Important Notes**
> Appear like this.

Get in touch

Feedback from our readers is always welcome.

General feedback: If you have questions about any aspect of this book, mention the book title in the subject of your message and email us at `customercare@packtpub.com`.

Errata: Although we have taken every care to ensure the accuracy of our content, mistakes do happen. If you have found a mistake in this book, we would be grateful if you would report this to us. Please visit `www.packtpub.com/support/errata`, select your book, click on the Errata Submission Form link, and enter the details.

Piracy: If you come across any illegal copies of our works in any form on the internet, we would be grateful if you would provide us with the location address or website name. Please contact us at `copyright@packt.com` with a link to the material.

If you are interested in becoming an author: If there is a topic that you have expertise in and you are interested in either writing or contributing to a book, please visit `authors.packtpub.com`.

Share Your Thoughts

Once you've read *OpenShift Multi-Cluster Management Handbook*, we'd love to hear your thoughts! Scan the QR code below to go straight to the Amazon review page for this book and share your feedback.

`https://packt.link/r/1803235284`

Your review is important to us and the tech community and will help us make sure we're delivering excellent quality content.

Download a free PDF copy of this book

Thanks for purchasing this book!

Do you like to read on the go but are unable to carry your print books everywhere?

Is your eBook purchase not compatible with the device of your choice?

Don't worry, now with every Packt book you get a DRM-free PDF version of that book at no cost.

Read anywhere, any place, on any device. Search, copy, and paste code from your favorite technical books directly into your application.

The perks don't stop there, you can get exclusive access to discounts, newsletters, and great free content in your inbox daily

Follow these simple steps to get the benefits:

1. Scan the QR code or visit the link below

https://packt.link/free-ebook/9781803235288

2. Submit your proof of purchase
3. That's it! We'll send your free PDF and other benefits to your email directly

Part 1 – Design Architectures for Red Hat OpenShift

This part is an introduction to the hybrid cloud journey and dilemma. Dive into the most important architecture concepts and definitions that should be done before deploying one or more OpenShift clusters.

This part of the book comprises the following chapters:

- *Chapter 1, Hybrid Cloud Journey and Strategies*
- *Chapter 2, Architecture Overview and Definitions*
- *Chapter 3, Multi-Tenant Considerations*
- *Chapter 4, OpenShift Personas and Skillsets*

1

Hybrid Cloud Journey and Strategies

Do you want to learn about and operate OpenShift in multiple environments? If you are reading this book, we suppose that the answer is yes! But before we go into the technical details, we want to start this book by making you a proposition: any house construction starts with a foundation, right? In this book, our approach will be the same. We will start by giving you the foundation to understand and create a much stronger knowledge base – you will develop critical thinking and be able to make the best decisions for your use case.

That is why we decided to start this book by not talking about OpenShift itself yet, but by unveiling the most popular (and important!) context that it operates in: **the hybrid cloud infrastructure**. Therefore, in this chapter, you will be introduced to the hybrid cloud journey, challenges, dilemmas, and why many organizations are struggling with it. Knowing about these challenges from the beginning is a determinant **success factor for hybrid cloud adoption**.

Transforming an IT business so that it's agile and scalable but is also stable is a must-have nowadays, but that is not a simple step; instead, it is a *journey from one star to another* in the vast outer space of IT that currently surrounds us. However, why do we need those changes? Why is the market adopting the cloud massively and so rapidly? We'll discuss that shortly!

It is a changing world!

We are living in the age of fast changes! 10 years ago, most of the current big tech companies did not exist or were just small startups; several technologies that we have today also were only known within university research groups, such as 3D printing, artificial intelligence and machine learning, 5G, edge computing, and others – and there is much more to come! Technologies like the ones mentioned previously are becoming more popular and will create several demands that do not exist today, new job positions, and far more changes.

In this rapidly changing world, some new needs became important. Most companies were forced to change to be able to release new software and versions much faster than before, quickly scale resources, and have a global presence with responsive applications.

It was in this context that the public cloud providers have emerged with great success. However, several organizations that made huge investments in the cloud are experiencing some challenges. In a study conducted by *IDG* in 2020, among big companies in different industries and geographies, 40% of the respondents stated that controlling cloud costs is the biggest challenge when taking full advantage of it. This research has also shown data privacy and security as big obstacles. We will walk through some of these challenges in this chapter.

In this chapter, we will cover the following topics:

- Main challenges of the public cloud
- How a hybrid cloud strategy helps mitigate these challenges
- How containers, Kubernetes, and OpenShift help to implement the hybrid cloud
- OpenShift options
- Types of OpenShift installations
- Additional tools to support hybrid cloud adoption

Main challenges of the public cloud

From small enterprises to big tech companies, most of them face some common challenges when it comes to using and taking full advantage of public cloud providers. Some of the main challenges are as follows:

- **Keeping cloud costs under control**: Estimating and managing the costs of applications running in a public cloud provider is not a simple thing – cloud providers' billing models are multifaceted, with hundreds of different options and combinations, each with a pricing factor. Finding the best cost-benefit for one application can take a significant amount of time. To make things even more complex, cloud costs are usually dynamic and flexible – this may change significantly from time to time by type, duration of the contract, type of computing resources, and so on.

- **Security**: Data privacy and security is one of the major concerns with public clouds, according to the *IDG's* research – almost 40% of them classified it as the top challenge. That is, it is naturally much more difficult to secure an IT environment that comprises of multiple different providers than the old days, in which the IT department usually only had a few on-premises environments to manage.

- **Governance, compliance, and configuration management**: Multiple providers mean different offerings and standards, probably different teams working with each of them, and, consequently, heterogeneous environments.

- **Integration**: Organizations that have legacy services and want to integrate with their applications, which are hosted in the cloud, usually face some dilemmas on the best way to do those integrations. While cloud providers virtually have no limits, when you integrate your applications with your legacy infrastructure, you might be creating a **harmful** dependency, which will limit their scalability. However, mainly for big enterprises, those integrations are inevitable, so how can we prevent dependency issues (or at least minimize them)?

- **Vendor lock-in**: A common concern when adopting cloud providers is often related to being locked in with a single vendor and the business risks associated with it. I would say that there is a thin line between getting the best price from the cloud provider and being locked into their services. What could happen to the business if the cloud provider decides to raise prices in the next contractual negotiation? Is this a risk your business can afford? How can we mitigate it? Here, the quote *you get what you pay for* is suitable!

- **Human resources and enablement**: Hiring and keeping talented people in IT has always been a hard task; cloud technologies are no different. Cloud engineer, Architect, SRE, Cloud Native Application Developer – these are just a few job positions that open every day, and most companies struggle to fill them. Hiring, training, and maintaining a skilled team to develop and operate applications in the cloud is a real challenge.

> **Reference**
>
> You can check out the complete *IDG* research at `https://www.idg.com/tools-for-marketers/2020-cloud-computing-study/` [Accessed 30 August 2021].

Benefits of the public cloud

We have seen some complex challenges so far. So, you might be thinking, *so you don't like cloud providers and want to convince me to avoid them, right?*

No, of course not! I am sure that without the advent of cloud providers, several companies we use every day (and love!) simply would not exist! Let's point out the good parts, then:

- **Scalability**: Cloud providers can offer almost unlimited and on-demand compute resources.

- **Lower CAPEX**: You don't need to buy any hardware and equipment to start any operations – you can do that with just a few clicks.

- **Resilience and global presence**: Even small companies can distribute services globally among different Availability Zones and Regions.

- **Modern technologies**: Public cloud providers are always looking to bring new and modern offerings, which helps an organization to always be at the edge of the technology.

Is hybrid cloud the solution?

As we've already discussed, the public cloud, while it can solve some challenges, introduces others. It was in this context that the hybrid cloud emerged: to mitigate some of the challenges and take the best from each provider, from on-premises, private, or cloud providers. The *HashiCorp State of Cloud Strategy Survey*, which was made in 2021 with more than 3,200 technology practitioners, found that multi-cloud is already a reality. 76% of the respondents stated that they are using multiple cloud vendors, with expectations for this to rise to 86% by 2023.

> **Reference**
>
> You can check out the complete *HashiCorp* research at `https://www.hashicorp.com/state-of-the-cloud` [Accessed 31 August 2021].

So, what are the characteristics of the hybrid cloud that help mitigate the challenges of public cloud adoption? Here are a few of them:

- Best-of-breed cloud services from different vendors can be combined, enabling a company to choose the best option for each workload.

- The ability to migrate workloads between different public and private cloud environments, depending on the actual circumstances.

- Being able to have a single, unified orchestration and management across all the environments for all providers.

The following table lists some of the challenges and hybrid cloud mitigations:

Challenge	Hybrid Cloud Mitigation
Cloud costs	Migrating workloads between different environments enables a company to use the best cost-benefit option for each case.
Security	The ability to integrate public and private environments allows you to choose the most secure option for any case.
Governance, compliance, and configuration management	Hybrid cloud adoption will drive an organization to establish a unified layer of orchestration and management between different providers. This will also help ensure compliance between the different environments.
Integration	Hybrid cloud means that the different providers will be integrated somehow – there are multiple ways to do this.
Vendor lock-in	The hybrid cloud strategy will give users the flexibility to make the best choices and move between different providers if needed; they won't be locked into one provider.
Human resources and enablement	Enablement will always be a challenge for technology. However, establishing a unified layer of orchestration and management between the cloud providers makes an organization less dependent on the knowledge from each cloud provider.

Containers and Kubernetes – part of the answer!

Containers have successfully emerged as one of the most important tools to promote better flexibility between applications and infrastructure. A container can encapsulate applications dependencies within a container image, which helps an application be easily portable between different environments. Due to that, containers are important instruments for enabling the hybrid cloud, although they have several other applications.

The following diagram shows how a container differs from traditional VMs in this matter:

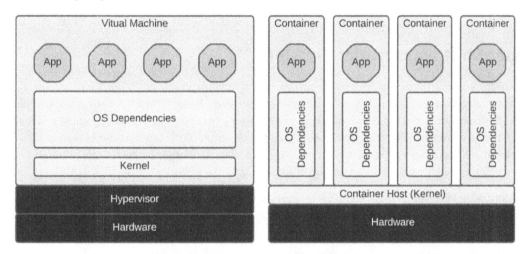

Figure 1.1 – Containers provide flexibility

While containers are beneficial, it is practically impossible to manage a large environment consisting of hundreds or thousands of containers without an orchestration layer. Kubernetes became the norm and it is a great orchestration tool. However, it is not simple to use. According to the *CNCF Survey 2020*, 41% of respondents see complexity as the top barrier for container adoption. When you decide to go for a vanilla Kubernetes implementation, some of the following will need to be defined (among a large set of options) and managed by you:

- Installation and OS setup, including configuration management
- Upgrades
- Security access and identity
- Monitoring and alerts
- Storage and persistence

- Egress, ingress, and network-related options

- Image scanning and security patches

- Aggregated logging tools

Reference

You can check out the complete *CNCF Survey* here: `https://www.cncf.io/blog/2020/11/17/cloud-native-survey-2020-containers-in-production-jump-300-from-our-first-survey/` [Accessed 1 September 2021].

OpenShift – a complete option

OpenShift is one of the most popular platforms based on Kubernetes among enterprise customers. It was first released in 2011, even before Kubernetes was created. However, in 2015, with the release of OpenShift version 3, Red Hat decided to adopt Kubernetes as its container orchestration layer. Since then, they are actively collaborating with the Kubernetes community – Red Hat and Google are the top contributors to Kubernetes. Due to that, it is not a surprise that OpenShift is one of the most mature and complete solutions built on top of Kubernetes.

The following table summarizes some of the features that are included out-of-the-box with the Red Hat **OpenShift Container Platform (OCP)** (or easily pluggable):

Feature	OpenShift Container Platform (OCP) Functionality
Built-in CI/CD Pipelines, Application Console	OpenShift pipelines(*), OpenShift GitOps(*), Developer Console.
Integrated Development Environment	OpenShift CodeReady Workspaces(*) and IDE extensions (VS Code and IntelliJ).
Serverless Middleware	OpenShift Serverless(*).
Service Mesh	OpenShift Service Mesh(*).
Automated Container Builds	S2I, BuildConfig.
Dashboard	Administrator and Developer dashboards are available.

Feature	OpenShift Container Platform (OCP) Functionality
Monitoring	Prometheus and Grafana.
Log Aggregation	OpenShift Logging(*).
Integrated Container Registry	Included with the platform.
Marketplace	Throughout Red Hat's online marketplace (`http://marketplace.redhat.com/`) and also the Operatorhub (`http://operatorhub.io/`), there's a community-driven catalog of operators, which contains operators from Red Hat, certified partners, and also from the open source community.
Self-Service Catalog	Included with Developer and Administrator Console using templates, helm charts, builder images, devfiles, or operators.
Multi-Tenant Clusters	RBAC, ResourceQuotas, Limits, and NetworkPolicies.
Operating System	Support included that's provided by Red Hat.
Network	OpenvSwitch or OVN-Kubernetes; internal DNS; ingress controllers (using HAProxy); Multus (for multiple networks); SR-IOV; Egress IP.
Storage	Plugins included (in-tree) that integrate with several providers (AWS EBS, Azure Disk, NFS, VMware vSphere, and others). You can also integrate with certified storage vendors using a **Container Storage Interface (CSI)**.
Machine Management	Automated server provisioning and patching.
Windows Containers	Support for workers using Windows Server (*).
Migration Toolkit for Containers	Stateful workload migration between clusters(*).
Kata Containers	OpenShift sandboxed containers(*).
Kubevirt (VMs)	OpenShift virtualization(*).
Over-the-Air Cluster Upgrades	Simplified upgrades of the entire stack (platform and OS) using the OTA feature.

(*) Need to be installed on day 2

These features are available for any customer that has a valid OpenShift subscription with Red Hat. However, if you don't have access to a Red Hat subscription, there are some alternatives (for studying purposes):

- You can use some of the trial options provided by Red Hat – check them out at `https://www.redhat.com/en/technologies/cloud-computing/openshift/try-it`.

- Use okd (`http://okd.io/`), which is the community distribution of OpenShift, also powered by Red Hat.

- Use Red Hat CodeReady Container in a VM on your desktop (requires an account on Red Hat's portal). More information can be found at `https://developers.redhat.com/products/codeready-containers/overview`.

We are going to see many of these great features in detail, along with practical examples, in this book.

> **Reference**
>
> The updated statistics about the contributions to the Kubernetes project, grouped by companies, can be found at `https://k8s.devstats.cncf.io/d/9/companies-table`.

OpenShift offerings – multiple options to meet any needs

An interesting factor about OpenShift is the vast range of platforms that are supported. With OpenShift version 4.11 (the version that was available when this book was written), you can have the following different combinations to choose from:

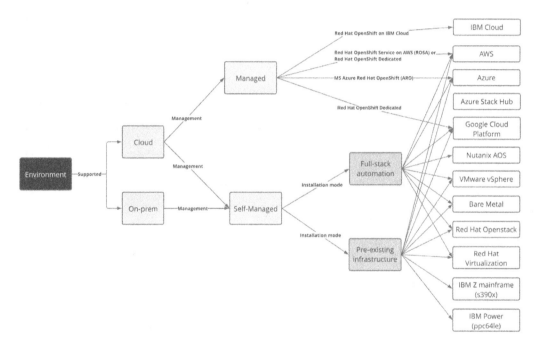

Figure. 1.2 – OpenShift offerings

In this section, we will walk through each of these options.

OpenShift managed cloud services

In the old days, when we talked about using a certain technology, we also thought about how to deploy and manage it. Nowadays, this is not always true – almost everything now can be found in a Software as a Service model, which you can quickly and easily start using without caring about deployment and management.

The same applies to OpenShift: multiple managed cloud services allow an organization to focus on the application's development and the business while Red Hat and the cloud provider manage the rest.

The following table shows the existing managed offerings at the time of writing this book (check Red Hat for the current options):

Offering Name	Cloud Provider	Billed By	Managed By	Supported By
Microsoft Azure Red Hat OpenShift	Azure	Microsoft	Red Hat and Microsoft	Red Hat and Microsoft
Red Hat OpenShift Dedicated	AWS or Google Cloud	1. Red Hat for OpenShift 2. AWS or Google Cloud for infrastructure used	Red Hat	Red Hat
Red Hat OpenShift on IBM Cloud	IBM Cloud	IBM	IBM	Red Hat and IBM
Red Hat OpenShift Service on AWS	AWS	AWS	Red Hat and AWS	Red Hat and AWS

Important Note

Note that Red Hat manages the full stack, not only the Kubernetes control plane. Red Hat provides management and version maintenance for the entire cluster, including masters, infrastructure, and worker nodes, though it's not limited to that: it also supports CI/CD, logging, metrics, and others.

There are other managed Kubernetes options on the market. Although this is not the focus of this book, keep in mind that some providers don't manage and support the entire stack – only the control plane, for instance. When you're considering a Kubernetes managed solution, see if it is fully managed or only part of the stack.

Managed or self-managed – which is the best?

The answer is: it depends! There are several things you need to consider to find out the best for your case, but generally speaking, managed solutions are *not* the best option for organizations that need to have control over the servers and their infrastructure. For organizations that are more focused on application development and don't care about the platform, as long as it is safe and reliable, then managed solutions are probably a good fit.

Managed solutions could also be helpful for organizations that want to put their hands on the platform, evaluate it, and understand if it fits their needs but don't have skilled people to maintain it yet.

Most of this book has been written with a self-managed cluster in mind. However, excluding the chapters focused on platform deployment and troubleshooting, the rest of it will likely apply to any type of OpenShift cluster.

The following diagram shows a workflow that aims to help you decide which strategy to go for:

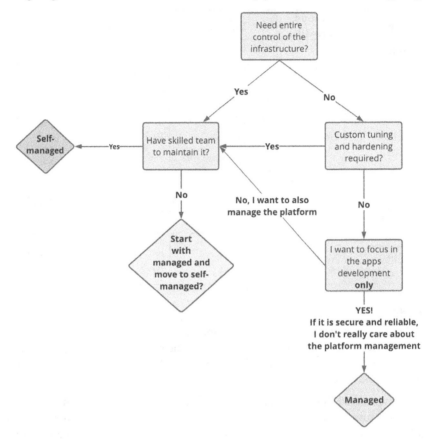

Figure. 1.3 – Managed or self-managed decision workflow

OpenShift installation modes

There are three installation modes you can use to deploy OpenShift in any of the supported providers, as follows:

- **Full-stack automated** (*installer-provisioned infrastructure*): In this mode, the installer will spin up all the required infrastructure automatically – the installer will integrate with the underlying virtualization or cloud provider to deploy all the machines that are required for the cluster. It is an opinionated fully automated solution that makes the deployment process a lot easier.

- **Pre-existing infrastructure** (*user-provisioned infrastructure*): With this installation, the machines are provisioned manually by following some standard images and processes, on top of tested virtualization or cloud providers.

- **Provider-agnostic** (also known as the *bare metal install method*): OpenShift is supported wherever Red Hat Enterprise Linux (*) is, though this doesn't mean that the installer and platform are tested (**) on every infrastructure layer combination that's supported with Red Hat Enterprise Linux. In such cases, you can use the provider-agnostic installation, which is a manual installation process with no integration between the installer and the platform with the virtualization or cloud provider.

(*) You can find a list of supported hypervisors for Red Hat Enterprise Linux at `https://access.redhat.com/certified-hypervisors`.

(**) Please refer to this link for an updated list of tested providers and integrations with OpenShift: `https://access.redhat.com/articles/4128421`.

OpenShift multi-cluster tools – going above and beyond

When it comes to supporting your hybrid or multi-cloud strategy, other great tools provide single and unified management, security, and orchestration layers across all environments in all providers. We reserved the last part of this book to take a deep dive into those tools, but you must meet them from the beginning to understand the role of each in the hybrid/multi-cloud picture.

Red Hat Advanced Cluster Management for Kubernetes – unified management

As we mentioned previously, a single and unified management layer is important to support the hybrid/multi-cloud strategy. Red Hat Advanced Cluster Management lets us manage the life cycle, ensure compliance using policies, and deploy applications on multiple Kubernetes clusters. The following are some of its main features:

- **Unified management**: Create, update, and delete Kubernetes clusters on top of different cloud providers. You can also access, find, and modify Kubernetes resources across the different clusters.

- **Governance, risk, and compliance**: Ensure compliance among multiple clusters using policies. Look for policy violations quickly and remediate them accordingly.

- **Application life cycle management**: Deploy applications across multiple clusters at once. Deploy complex applications by integrating Advanced Cluster Management with Red Hat Ansible Automation Platform to configure networks, load balancers, and other external dependencies.

- **Multi-cluster observability**: Check the health status of multiple clusters from a single point using out-of-the-box dashboards and metrics.

We will dive into Red Hat Advanced Cluster Management using practical examples in the last part of this book.

Red Hat Advanced Cluster Security for Kubernetes – securing applications no matter where they are

Security is becoming increasingly important for Kubernetes users. When you have multiple Kubernetes clusters spread among different providers, ensuring security and having a real notion of the current vulnerabilities is a real challenge. Red Hat Advanced Cluster Security aims to help with that – through it, you can easily *scan container images to find known vulnerabilities, audit workloads, and clusters using industry standards such as NIST, PCI, and others, analyze network traffic, and create policies accordingly, among other great features.* You can apply all of these features to multiple different clusters, which helps you keep all your environments secure, no matter where they are.

We will look at Red Hat Advanced Cluster Security using practical examples in the last part of this book.

Red Hat Quay – storing and managing container images in a central repository

A central container image registry isn't usually a required tool. However, deploying applications on several clusters without it makes the build and deployment activity a bit challenging. Red Hat Quay is a container image registry that provides not only the usual capabilities of an image registry (storing your container images) but also provides image vulnerability scans, a time machine, replication, garbage collection, automated builds, authentication, authorization, and more.

We will learn how to use Red Hat Quay in the last part of this book.

OpenShift Plus – the whole package

Red Hat OpenShift, Advanced Cluster Management, Advanced Cluster Security, and Quay are different products. However, with the OpenShift Plus package, you can have all of them in one subscription only, which is probably the best way to go if you are planning to adopt a hybrid or multi-cloud Kubernetes strategy.

We will cover OpenShift Plus in more detail with practical examples in the last part of this book.

Summary

In this chapter, we looked at the main challenges of public cloud usage and how the hybrid cloud helps mitigate some of them. You now understand how containers, Kubernetes, and OpenShift can help you implement a successful hybrid cloud strategy. Finally, we learned about the different types of OpenShift offerings and additional tools that support hybrid cloud adoption.

In the next chapter, you will learn about the architectural aspects of an OpenShift cluster.

Further reading

If you want to find out more about the concepts that were covered in this chapter, check out the following references:

- The following are the public studies and surveys that were mentioned in this chapter related to hybrid cloud and container adoption:

  - *IDG 2020 Cloud Computing Study:* `https://www.idg.com/tools-for-marketers/2020-cloud-computing-study/`

  - *HashiCorp State of Cloud Strategy Survey:* `https://www.hashicorp.com/state-of-the-cloud`

  - *Cloud-Native Survey 2020: Containers in production jump up 300% from our first survey:* `https://www.cncf.io/blog/2020/11/17/cloud-native-survey-2020-containers-in-production-jump-300-from-our-first-survey/`

- *Red Hat OpenShift landing page:* `https://www.redhat.com/en/technologies/cloud-computing/openshift`

- *Red Hat Managed cloud services landing page:* `https://www.redhat.com/en/technologies/cloud-computing/openshift/managed-cloud-services`

- *Red Hat Advanced Cluster Security for Kubernetes landing page:* `https://www.redhat.com/en/technologies/cloud-computing/openshift/advanced-cluster-security-kubernetes`

- *Red Hat Advanced Cluster Management for Kubernetes landing page:* `https://www.redhat.com/en/technologies/management/advanced-cluster-management`

- *Red Hat Quay landing page:* `https://www.redhat.com/en/technologies/cloud-computing/quay`

- *Red Hat OpenShift Platform Plus landing page:* `https://www.redhat.com/en/technologies/cloud-computing/openshift/platform-plus`

- *OpenShift Container Platform 4.x Tested Integrations and Supportability Matrix:* `https://access.redhat.com/articles/4128421`

- *OpenShift Container Platform installation overview (from official documentation):* `https://docs.openshift.com/container-platform/latest/installing/index.html`

- *Supported installation methods for different platforms (from official documentation):* `https://docs.openshift.com/container-platform/latest/installing/installing-preparing.html#supported-installation-methods-for-different-platforms`

- *Kubernetes statistics page:* `https://k8s.devstats.cncf.io/`

2
Architecture Overview and Definitions

Kubernetes is an amazing technology; however, as we saw in the last chapter, it is not a simple technology. I consider Kubernetes not only as container orchestration, but besides that, it is also a platform with standard interfaces to integrate containers with the broader infrastructure, including storage, networks, and hypervisors. That said, you must consider all the prerequisites and aspects involved in an OpenShift self-managed cluster.

In this chapter, we will walk through the main concepts related to the Kubernetes and OpenShift architecture. The main purpose here is you *think before doing* and make important decisions, to avoid rework later.

The following main topics will be covered in the chapter:

- Understanding the foundational concepts
- OpenShift architectural concepts and best practices
- Infrastructure/cloud provider
- Network considerations
- Other considerations
- OpenShift architectural checklists

Let's get started!

Technical requirements

As we are covering the architecture side of OpenShift in this chapter, you still don't need access to any specific hardware or software to follow this chapter, but this will be expected some chapters ahead. However, it is important you have some pre-existing knowledge of OpenShift and Kubernetes for you to achieve the best possible result from this chapter.

Prerequisites

This chapter is intended to be for **Information Technology** (IT) architects that already have some basic knowledge of Kubernetes or OpenShift use. That said, we are not covering in this chapter basic concepts such as what a Pod, Service, or Persistent Volume is. But if you don't know these basic concepts yet, don't freak out! We have prepared a list of recommended training and references for you in the last chapter of this book. We suggest you to take the Kubernetes Basics and Kube by Example before moving forward with this chapter.

Understanding the foundational concepts

Let's start by understanding the main concepts related to Kubernetes and OpenShift components and servers. First, any OpenShift cluster is composed of two types of servers: **master** and **worker** nodes.

Master nodes

This server contains the **control plane** of a Kubernetes cluster. Master servers on OpenShift run over the **Red Hat Enterprise Linux CoreOS** (**RHCOS**) operating system and are composed of several main components, such as the following:

- **Application programming interface (API) server** (`kube-apiserver`): Responsible for exposing all Kubernetes APIs. All actions performed on a Kubernetes cluster are done through an API call—whenever you use the **command-line interface** (**CLI**) or a **user interface** (**UI**), an API call will always be used.

- **Database** (`etcd`): The database stores all cluster data. `etcd` is a highly available distributed key-value database. For in-depth information about `etcd`, refer to its documentation here: `https://etcd.io/docs/latest/`.

- **Scheduler** (`kube-scheduler`): It is the responsibility of `kube-scheduler` to assign a node for a Pod to run over. It uses complex algorithms that consider a large set of aspects to decide which is the best node to host the Pod, such as computing resource available versus required node selectors, affinity and anti-affinity rules, and others.

- **Controller manager** (`kube-controller-manager`): Controllers are an endless loop that works to ensure that an object is always in the desired state. As an illustration, think about smart home automation equipment: a controller is responsible for orchestrating the equipment to make sure the environment will always be in the desired programmed state—for example, by turning the air conditioning on and off from time to time to keep the temperature as close as possible to the desired state. Kubernetes controllers have the same function— they are responsible for monitoring objects and responding accordingly to keep them at the desired states. There are a bunch of controllers that are used in a Kubernetes cluster, such as replication controller, endpoints controller, namespace controller, and serviceaccounts controller. For more information about controllers, check out this page: `https://kubernetes.io/docs/concepts/architecture/controller/`

These are the components of the Kubernetes control plane that runs on the master nodes; however, OpenShift has some additional services to extend Kubernetes functionality, as outlined here:

- **OpenShift API Server** (`openshift-apiserver`): This validates and configures data for OpenShift exclusive resources, such as routes, templates, and projects.

- **OpenShift controller manager** (`openshift-controler-manager`): This works to ensure that OpenShift exclusive resources reach the desired state.

- **OpenShift Open Authorization (OAuth) server and API** (`openshift-oauth-apiserver`): Responsible for validating and configuring data to authenticate a user, group, and token with OpenShift.

The following figure shows the main control plane components of Kubernetes and OpenShift:

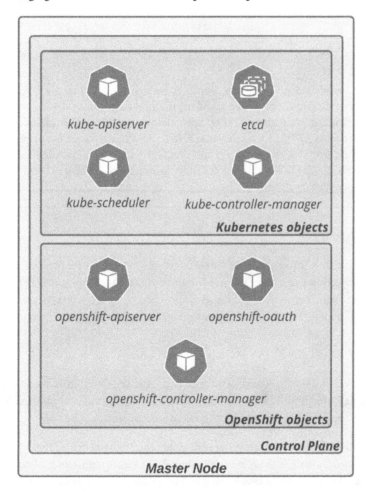

Figure 2.1 – OpenShift control plane components

These components can be found in multiple namespaces, as you can see in the following table:

Control plane component	Namespace	Managed by
kube-apiserver	openshift-kube-apiserver	kube-apiserver operator
Etcd	openshift-etcd	etcd operator
kube-scheduler	openshift-kube-scheduler	kube-scheduler operator
kube-controller-manager	openshift-kube-controller-manager	kube-controller-manager operator
openshift-apiserver	openshift-apiserver	openshift-apiserver operator
openshift-controller-manager	openshift-controller-manager	openshift-controller-manager operator
openshift-oauth	openshift-oauth-apiserver	Authentication operator

> **What Are Operators?**
>
> If you've never heard about Operators, you may be thinking: *What are Operators, after all?* Operators are nothing more than a standard method to package, deploy, and maintain Kubernetes applications and objects. They use **Custom Resource Definitions** (**CRDs**) to extend the Kubernetes API functionality and also some standards for the application's life cycle: deploy, patch, keep the desired state, and even auto-pilot it (autoscaling, tuning, failure detections, and so on). Check out this link for more information: `https://kubernetes.io/docs/concepts/extend-kubernetes/operator/`.

Bootstrap node

The bootstrap node is a temporary server—used only during cluster deployment—that is responsible for injecting the cluster's components into the control plane. It is removed by the installation program when the bootstrap is finished successfully. As it is a temporary server that lives only during the deployment, it is usually not considered in the OpenShift architecture.

Workers

Workers are the servers where the workloads themselves run. On OpenShift, workers can run over RHCOS or **Red Hat Enterprise Linux** (**RHEL**). Although RHEL is also supported for OpenShift workers, RHCOS, in general, is preferred for the following reasons:

- **Immutable**: RHCOS is a tight operating system designed to be managed remotely by OpenShift Container Platform itself. This enables consistency and makes upgrades a much easier and safer procedure, as OpenShift will always know and manage the actual and desired states of the servers.

- `rpm-ostree`: RHCOS uses the `rpm-ostree` system, which enables transactional upgrades and adds consistency to the infrastructure. Check out this link for more information: `https://coreos.github.io/rpm-ostree/`.

- **CRI-O container runtime and container tools**: RHCOS's default container runtime is **CRI-O**, which is optimized for Kubernetes (see `https://cri-o.io/`). It also comes with a set of tools to work with containers, such as `podman` and `skopeo`. During normal functioning, you are not encouraged to access and run commands on workers directly (as they are managed by the OpenShift platform itself); however, those tools are helpful for troubleshooting purposes—as we will see in detail in *Chapter 6* of this book, *OpenShift Troubleshooting, Performance, and Best Practices*.

- **Based on RHEL**: RHCOS is based on RHEL—it uses the same well-known and safe RHEL kernel with some services managed by `systemd`, which ensures the same level of security and quality you would have by using the standard RHEL operating system.

- **Managed by Machine Config Operator (MCO)**: To allow a high level of automation and also keep secure upgrades, OpenShift uses the MCO to manage the configurations of the operating system. It uses the `rpm-ostree` system to make atomic upgrades, which allows safer and easier upgrade and rollback (if needed).

In the following figure, you can view how these objects and concepts are used in an OpenShift worker node:

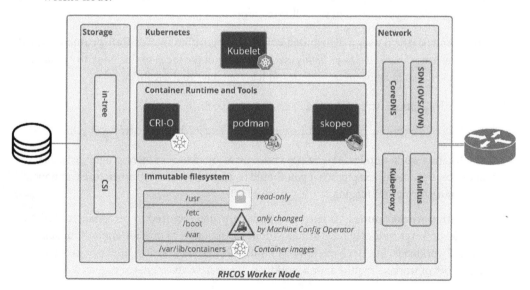

Figure 2.2 – RHCOS worker node

Types of workers

There are some common types of workers, the most usual being these:

- **Application workers**: Responsible for hosting the workloads—this is where the application containers run.

- **Infrastructure workers**: This type of server is usually used to host the platform infrastructure tools, such as the ingress (routers), internal registry, the monitoring stack (Prometheus and Grafana), and also the logging tool (Elasticsearch and Kibana).

- **Storage workers**: Container storage solutions, such as **Red Hat OpenShift Data Foundation**, usually require some dedicated worker nodes to host their Pods. In such cases, a best practice is to use a dedicated node group for them.

In the next section, we will see how to use different types of workers to design a highly available and resilient OpenShift cluster.

Highly available cluster

It is not uncommon for OpenShift clusters to become critical for the enterprise—sometimes, they start small but become large really quickly. Due to that, you should consider in your OpenShift cluster architecture **non-functional requirements (NFRs)** such as **high availability (HA)** from day one. A highly available cluster is comprised of the following aspects:

- **Master nodes**: `etcd` uses a distributed consensus algorithm named **Raft protocol**, which requires at least *three nodes to be highly available*. It is not the focus of this book to explain the Raft protocol, but if you want to understand it better, refer to these links:

  - Raft description: `https://raft.github.io/`

  - Illustrated example: `http://thesecretlivesofdata.com/raft/`

- **Infrastructure worker nodes**: At least two nodes are required to have ingress highly available. We will discuss later in this chapter what you should consider for other infrastructure components such as monitoring and logging.

- **Application worker nodes**: At least two nodes are also required to be considered highly available; however, you may have as many nodes as required to provide enough capacity for expected workloads. In this chapter, we will walk through some sizing guidance to determine the number of workers required for a workload, if you have an estimated required capacity.

The following figure shows what a highly available cluster architecture looks like:

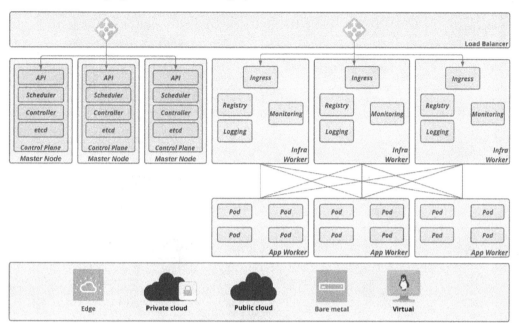

Figure 2.3 – OpenShift highly available cluster

Now that we are on board with the foundational concepts of Kubernetes and OpenShift, let's dive further and look at OpenShift's architecture, along with some best practices.

OpenShift architectural concepts and best practices

In this section, we will discuss the main concepts related to the OpenShift architecture design and some best practices you should consider. In general, when we are designing an architecture for an OpenShift cluster, the aspects in the following table need to be defined accordingly:

Aspect	Item	Description
Infrastructure or cloud provider	Define a provider	Once you have decided which provider your cluster will be hosted on, keep in mind the requirements you have with the infrastructure or cloud provider, depending on the installer mode defined.
Installation mode	**User-provisioned infrastructure (UPI)**, **installer-provisioned infrastructure (IPI)**, or agnostic installer?	As we already covered in the last chapter, there are multiple installation modes. Define what type of installer you will use.
Infrastructure or cloud provider	User permissions	Once you have decided which provider and installation mode you will use, keep in mind the requirements you have to provide in terms of the infrastructure or cloud provider.
Computing	Cluster sizing	Define the number and size of workers to host desired applications.
Aggregated logging tool	Use OpenShift Logging or an external solution?	Define whether you are going to use the OpenShift out-of-the-box logging solution or an external solution.
Monitoring	Which monitoring solution to use	You need to decide which monitoring solution to go for: the native OpenShift monitoring solution or another tool.
Storage	Define storage backend	We are going to see that there are many options to provide persistence to your containers on OpenShift. You need to determine which types of volumes you are going to use and what will be the storage backend for each.
Storage	Sizing	As you have decided which storage backend you will use with your OpenShift cluster, it is important to also estimate the initial amount of storage that will be required and ensure that the storage backend will have enough capacity to provide that.
Network	Define network	OpenShift supports the installation of a cluster in an existing **virtual private cloud (VPC)** (on cloud providers) or a new one. If you decide to go for an existing one, you will also need to define and create subnets and other network-related things manually.
Network	Define **software-defined networking (SDN)** subnet ranges	We will see in this chapter that OpenShift SDN uses two internal subnets. You need to carefully review it and ensure it will not conflict with any existing real subnet in your infrastructure.
Network	Configure **Domain Name System (DNS)**	You need to observe the DNS requirements, which will be different between UPI and IPI deployments.
Network (on-premises)	Define external load balancer	OpenShift on-premises already has an internal highly available load-balancing mechanism. However, depending on the load expected, you may decide to go for an external load balancer appliance.
Network (on-premises)	**Dynamic Host Configuration Protocol (DHCP)/Intelligent Platform Management Interface (IPMI)/Preboot eXecution Environment (PXE)**	For UPI and bare-metal installations, other requirements need to be observed.
Proxy/firewall	Using a firewall, proxy, or restricted network	The most convenient way to install OpenShift is when connected to Red Hat registries over the internet; however, it is possible to deploy it on restricted networks.
Secure Sockets Layer (SSL) certificates	Custom or self-signed?	You may use self-signed certificates or change them to use your custom certificates.
Identity provider (IdP) (authentication)	HTPasswd, **Lightweight Directory Access Protocol (LDAP)**, OpenID, GitHub, and so on	Define which will be the authentication mechanism used with the cluster.

Details of how to address these aspects (deployment, configurations, and so on) will be covered from *Chapter 5, OpenShift Deployment,* onward. Another key point to note is that we are still focusing on one single OpenShift cluster only—the main objective here is to help you to define a standard cluster architecture that best fits your case. In the next chapter, we will explore aspects you need to consider when working with multiple environments, clusters, and providers.

In the following sections, we are going to walk through each of these points, highlighting the most important items you need to cover.

Installation mode

You already know from the previous chapter that we have three installation modes with OpenShift, as follows:

- **Installer-provisioned infrastructure (IPI)**

- **User-provisioned infrastructure (UPI)**

- Provider-agnostic (if you haven't seen it, review the *OpenShift installation modes* section from the last chapter)

Here, we will briefly discuss important aspects you need to consider from each option to drive the best decision for your case.

IPI

This mode is a simplified opinionated method for cluster provisioning and is also a fully automated method for installation and upgrades. With this model, you can make the operational overhead lower; however, it is less flexible than UPI.

You can see an example of IPI here:

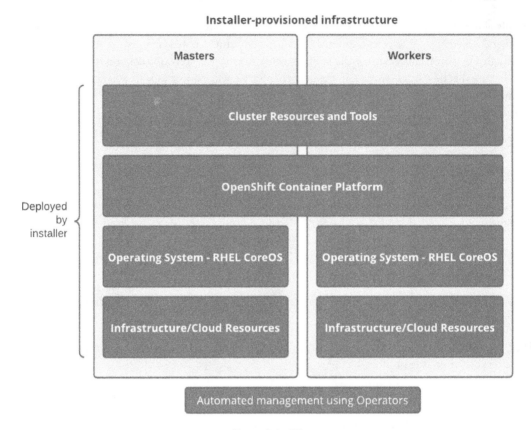

Figure 2.4 – IPI

Figure 2.4 shows all layers that are automated by the installer during the cluster deployment.

UPI

In this mode, you provision the servers manually—you are also responsible for managing them. As such, you have more flexibility within the infrastructure layer. In this mode, OpenShift still has some level of integration with the infrastructure or cloud provider to provide storage services for the platform.

Agnostic installer

This mode is similar to UPI; however, there is no integration between OpenShift and the infrastructure or cloud provider. Therefore, in this mode, you will not have any storage plugins installed with the platform—you will need to deploy an in-tree or **Container Storage Interface** (**CSI**) plugin on day two to provide persistent volumes to your workloads (we are going to cover storage-related aspects later in this chapter).

You can see an example of UPI/an agnostic installer here:

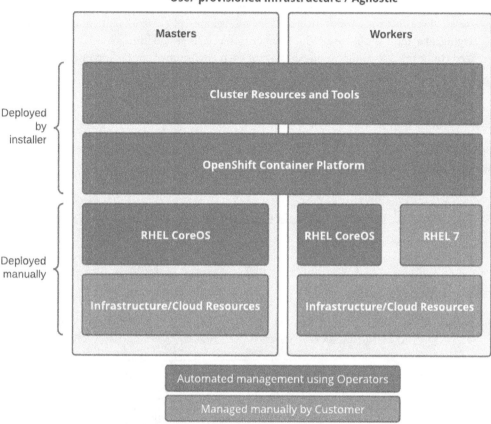

Figure 2.5 – UPI/Agnostic installer

As you can see in *Figure 2.5*, with UPI or an agnostic installer, there are some layers you are responsible for providing, as prerequisites, to deploy a cluster (and also maintain it on day two), as opposed to IPI, from *Figure 2.4*, which is fully automated.

Computing

From a computing perspective, the following are important attributes that must be considered during the architecture design:

- **Nodes and cluster sizing**: Define the number and size of worker nodes to host workloads expected for the platform. Some important factors need to be considered here to have a resilient cluster—this topic will be covered later in this chapter.

- **Environment segmentation**: It is possible to have one cluster only that provides a segregated group of nodes for specific reasons. Sometimes, it makes sense to have a dedicated group of nodes to provide services for specific environments in one single cluster—it is possible to have one single cluster with nodes dedicated for a development environment, another group for staging, and another one for production, for instance. That said, this is a crucial decision that needs to be made—going for one cluster for each environment or having one single cluster that serves multiple environments. We are going to explore this point in the next chapter and see what the pros and cons of each case are.

Master nodes' sizing

To define the master nodes' size, we recommend you follow Red Hat's benchmark, based on expected cluster load and number of nodes, as follows:

Number of worker nodes	Cluster load (namespaces)	Central processing unit (CPU) cores	Memory (gigabytes (GB))
25	500	4	16
100	1000	8	32
250	4000	16	96

Infrastructure node sizing

Similarly, infrastructure nodes' size also has a benchmark, based on expected cluster size, as follows:

Number of worker nodes	CPU cores	Memory (GB)
25	4	16
100	8	32
250	16	128
500	32	128

However, the preceding table does not consider OpenShift logging. Therefore, if you are planning to use it, add at least four more **virtual CPUs (vCPUs)** and 16 GB to the nodes on which Elasticsearch instances will be hosted.

Worker nodes' sizing

There isn't just one algorithm to estimate the size of an OpenShift cluster. The sizing algorithm we listed here is based on our personal experience along the years working with it, and also great articles and resources we have studied so far—some good references on this topic are available at the end of this chapter in the *Further reading* section.

Allocatable resources

The sizing estimation rationale for computing resources needs to consider the nodes' allocatable resources. The allocatable resource is the real amount that can be used for workloads in a node, considering the number of resources that are reserved for the operating system and `kubelet`. The calculation of allocatable resources is given by the following formula:

$$Allocatable\ Resources$$
$$= [Node\ Capacity] - [kube - reserved] - [system - reserved]$$
$$- [hard - eviction - threshold]$$

OpenShift Default Values

The default values for OpenShift workers are as follows (at the time of this writing):

CPU:

- `system-reserved = 500m (*)`

- `kube-reserved = 0m (*)`

- `hard-eviction = 0m (*)`

Memory:

- `system-reserved = 1Gi`

- `kube-reserved = 0Gi`

- `hard-eviction = 100Mi`

(*) "m" stands for *millicore*, a standard Kubernetes unit that represents one vCPU divided into 1,000 parts.

Recommended allocatable resources

Besides the standard aforementioned allocatable resources, it should also be considered as a best practice to keep at least 25% of resources available in a node, for resilience purposes. I'll explain: when one node goes down, the native Kubernetes resilience mechanism, after some time, will move the Pods to other nodes with available resources—that means if you don't plan to have extra capacity on the nodes, this resilience mechanism is at risk. You should also consider extra capacity for autoscaling at peak times and future growth. Therefore, it is recommended you consider this extra capacity in the calculation of workers' computing sizing, as follows:

$$Recommended\ Allocatable\ Resources = [Allocatable\ Resources] * 0.75$$

> **Important Note**
>
> Usually, some level of CPU overcommitment is—somewhat—handled well by the operating system. That said, the extra capacity mentioned previously doesn't always apply to the CPU. However, this is a workload-dependent characteristic: most container applications are more memory- than CPU-bound, meaning that CPU overcommitment will not have a great impact on overall application performance, while the same does not happen with memory—but again, check your application's requirement to understand that.

Let's use an example to make this sizing logic clear.

Example

Imagine that you use servers with 8 vCPUs and 32 GB random-access memory (RAM) as the default size. A worker of this size will have, in the end, the following recommended allocatable resources:

- CPU

• Formula	• Example
• $ar = [nc] - [kr] - [sr] - [he]$ • $rar = ar * 0.75$	• $ar = 8000 - 0 - 500 - 0 = 7500$mi $rar = 7500 * 0.75 = \mathbf{5625mi}$ with: • nc = 8000 • system-reserved = 500 (default) • kube-reserved = 0 (default) • hard-eviction-threshold = 0 (default)

- Memory:

• Formula	• Example
• $ar = [nc] - [kr] - [sr] - [he]$ • $rar = ar * \mathbf{0.75}$	• $ar = 32000 - 0 - 1000 - 100 = 30900$MB • $rar = 30900 * 0.75 = \mathbf{23175\ MB\ RAM}$ with: • nc = 32000 • system-reserved = 0 (default) • kube-reserved = 1000 (default) • hard-eviction-threshold = 100 (default)

Legend:

ar = allocatable resources

nc = node capacity

kr = kube-reserved

sr = system-reserved

he = hard-eviction threshold

Therefore, a worker with 8 vCPUs and 32 GB RAM will have approximately **5 vCPUs and 23 GB RAM** considered as the usable capacity for applications. Considering an example in which an application Pod requires on average 200 millicores and 1 GB RAM, a worker of this size would be able to host approximately 23 Pods (limited by memory).

Aggregated logging

You can optionally deploy the **OpenShift Logging** tool that is based on **Elasticsearch**, **Kibana**, and **Fluentd**. The following diagram explains how this tool works:

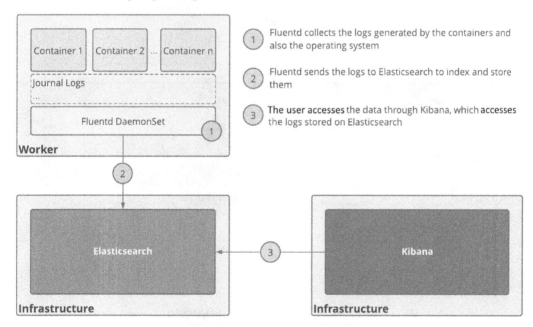

Figure 2.6 – OpenShift Logging components

You are not required to use OpenShift Logging, though, if you have your logging solution and want to keep it. You only need to configure the `ClusterLogForwarder`, as you are going to see in later chapters of this book (from *Chapter 5, OpenShift Deployment,* onward).

Monitoring

Another important tool that any container orchestration platform needs to have is a monitoring tool that can monitor your infrastructure and applications. OpenShift comes natively with a monitoring solution based on **Prometheus**, **AlertManager**, and **Grafana**. The following diagram explains the monitoring components:

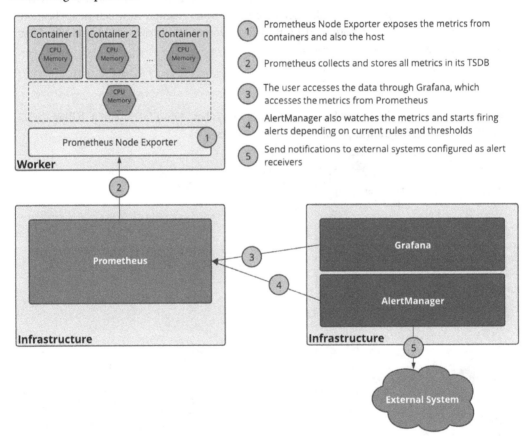

Figure 2.7 – OpenShift monitoring components

OpenShift monitoring is not optional; it is used by many internal platform components. However, if you do not intend to use it in favor of another monitoring tool, you may keep it using ephemeral storage. On the other hand, if you are planning to use it, we recommend you provide persistent storage to save the monitoring metrics.

Storage

Containers are stateless by nature, but this does not mean that it is not possible to have stateful containers on OpenShift. There are multiple ways to mount storage volumes inside containers and enable stateful workloads. In the following sections, we will walk through the common storage requirements of an OpenShift cluster that you should consider in your architectural design.

Storage backends

There are two types of storage implementations: in-tree and CSI plugins.

In-tree volume plugins

In-tree plugins are implementations that allow a Kubernetes platform to access and use external storage backends. The name *in-tree* comes from the fact that these implementations are developed and released in the main Kubernetes repositories, as *in-tree* modules. There are several types of supported in-tree plugins with OpenShift, as follows (*):

Volume plugin	ReadWriteOnce (RWO)	ReadOnlyMany (ROM)	ReadWriteMany (RWX)
AWS EBS	✓	-	-
Azure File	✓	✓	✓
Azure Disk	✓	-	-
Cinder	✓	-	-
Fibre Channel	✓	✓	-
GCE Persistent Disk	✓	-	-
HostPath	✓	-	-
iSCSI	✓	✓	-
Local volume	✓	-	-
NFS	✓	✓	✓
Red Hat OpenShift Container Storage	✓	-	✓
VMware vSphere	✓	-	-
emptyDir	✓	-	-

(*) At the time this book was written. Check the currently supported options at https://access.redhat.com/articles/4128421.

CSI drivers

With more and more storage providers supporting Kubernetes, the development and maintenance of in-tree plugins became difficult and was no longer the most efficient model. The CSI has been created in this context: to provide a standard way to extend Kubernetes storage capabilities using API interfaces—as such, you can easily add new CSI plugins for different storage providers and use them with OpenShift. With CSI, it is possible to also have interesting features such as **snapshots, resizing, and volume cloning**; however, it is up to the storage provider to implement these features or not, so check with them if they have a CSI driver implementation available and which operations are implemented and supported.

> **Important Note**
> Red Hat supports the CSI APIs and implementation from the OpenShift side; however, support of the storage side is a storage vendor's responsibility. Check with your storage vendor if there is a supported CSI option for OpenShift.

Storage requirements

Now that you have learned about the types of storage plugins available for OpenShift, let's review the storage requirements you usually have with an OpenShift cluster.

Server disks

OpenShift servers use one disk with 120 GB by default. Large clusters require master nodes with low latency and high throughput disks, which can provide at least 500 sequential **input/output operations per second (IOPS)** (usually **solid-state drive (SSD)** or **Non-Volatile Memory Express (NVMe)** disks). We are also going to see in-depth details about this in *Chapter 5, OpenShift Deployment*.

OpenShift internal registry

This depends on the number and size of application images to be stored on it. If you do not have an estimated value for the images, an initial size of 200 GB is usually enough for the first few weeks. As a best practice, consider setting image pruner policies to automatically delete images that are no longer used—we are going to cover these best practices with examples in *Chapter 5, OpenShift Deployment*.

Volume type used by OpenShift internal registry: **RWX**

OpenShift Logging

This depends on the number of logs generated by the applications. Here is an example of the volume size required for an application that generates 10 lines of logs per second (lines-per-second); the lines have 256 bytes (bytes-per-line) on average, considering a retention period of 7 days for the logs:

$$bytes\ per\ line - lines\ per\ second = 10 * 256 = 2560\ bytes\ per\ pod\ per\ second$$
$$= 2560 * 60 * 60 * 24 = 221{,}180{,}000\ bytes\ per\ day = 0.221184\ GB * 7$$
$$= \mathbf{1.548288\ GB\ per\ pod}$$

This means that one single Pod of that application will consume nearly 1.5 GB over 7 days (the period for which a log will be stored on Elasticsearch). Another important thing to consider is Elasticsearch's replication factor, which will require more storage depending on the replication factor selected. There following replication factors are available:

- `FullRedundancy`: Replicates the primary shards for each index to every Elasticsearch node
- `MultipleRedundancy`: Replicates the primary shards for each index to 50% of the Elasticsearch nodes
- `SingleRedundancy`: Makes one copy of the primary shards for each index
- `ZeroRedundancy`: Does not make a copy of the primary shards

Volume type used by OpenShift Logging: **RWO**

OpenShift monitoring

OpenShift monitoring is installed by default with the platform using ephemeral storage (also known as **emptyDir**), meaning that, for some reason, when the Prometheus pod gets restarted, all metrics data will be lost. To avoid losing metrics data, consider a persistent volume for **Prometheus** and **AlertManager** Pods.

Red Hat has a benchmark based on various tests performed, as represented here. This empirical data is good guidance to estimate the volume required for Prometheus:

Number of nodes	Number of pods	Prometheus storage growth per day	Prometheus storage growth per 15 days (*)	RAM space (per scale size)	Network (per tsdb chunk)
50	1800	6.3 GB	94 GB	6 GB	16 megabytes (MB)
100	3600	13 GB	195 GB	10 GB	26 MB
150	5400	19 GB	283 GB	12 GB	36 MB
200	7200	25 GB	375 GB	14 GB	46 MB

(*) 15 days is the default retention period.

You also need to consider volumes for **AlertManager**: typically, a volume size of **20 GB** is enough for most cases.

By default, an HA configuration is composed of **two Prometheus replicas and three AlertManager replicas**.

Using the preceding reference, we can estimate the volumes required for OpenShift monitoring. For example, let's say that we are planning a cluster that will have no more than 50 nodes and 1,800 Pods. In that case, we'd need to use the following formula:

$$Space\ required\ for\ Prometheus = 94\ GB\ (required\ for\ 15\ days) * 2\ pods = 188\ GB$$

$$Space\ required\ for\ AlertManager = 20\ GB * 3\ pods = 60\ GB$$

$$Total\ required\ for\ a\ cluster\ with\ 50\ nodes\ and\ 1800\ pods = 188 + 60 = \mathbf{248\ GB}$$

Volume type used by OpenShift monitoring: **RWO**

> **Note**
>
> The preceding requirements are based on empirical data. The real consumption observed can be higher depending on the workloads and resource usage. For more information, refer to the official documentation: `https://docs.openshift.com/container-platform/ latest/scalability_and_performance/scaling-cluster-monitoring- operator.html`.
>
> At this time, you don't need to know in-depth details about the OpenShift components such as logging or monitoring, as we are only covering the amount of storage required (or estimated) for them. These tools will be covered in detail later in this book.

Example

As we have already addressed sizing guidelines for an OpenShift cluster, let's use an example to make it clearer. Imagine that we are designing an OpenShift cluster architecture that is planned to host a three-tier Node.js application with the following capacity:

- Up to 20 Pods on the frontend consume 300 millicores and 1 GB RAM each at peak load. Each pod generates 30 lines of logs per second (256 bytes per line). Stateless Pods.

- Up to 4 Pods on the backend need 500 millicores and 1 GB RAM each at peak load. Each pod generates 10 lines of logs per second (256 bytes per line). Stateless Pods.

- 1 MongoDB database instance with 8 GB RAM and 2 vCPUs. It generates 1 line of logs per second (256 bytes per line). An **RWO** volume is required of 500 GB.

Our logging stack is configured with `ZeroRedundancy` (there is no data replication).

Compute sizing

First, let's see the total amount of CPU and memory required (for workloads only), as follows:

- CPU

- $Front-end = 20 * 300 = 6000\ millicores = 6\ vCPU$
  $Back-end = 4 * 500 = 2000\ millicores = 2\ vCPU$
  $Total = \mathbf{10\ vCPU}$

- Memory

- $Front-end = 20 * 1 = 20\ GB\ RAM$
  $Back-end = 4 * 1 = 4\ GB\ RAM$
  $MongoDB = 8\ GB\ RAM$
  $Total = \mathbf{32\ GB\ RAM}$

- Volume

- $MongoDB = \mathbf{500\ GB\ RWO}$

We will assume nodes with 4 vCPUs and 16 GB RAM by default. As we saw in this chapter, we need to apply the following formula to define the recommended allocatable resources:

- CPU

- $Allocatable\ Resources\ (millicore)$
  $= [Node\ Capacity] - [kube-reserved] - [system-reserved] - [hard-eviction-threshold] = 4000 - 0 - 500 - 0 = \mathbf{3500mi}$

> **Note**
>
> We are considering, in this case, that some level of CPU overcommit is acceptable, and due to that, we are not considering the 25% of extra capacity here (recommended allocatable resources).

- Memory

$Allocatable\ Resources\ (millicore)$
$= [Node\ Capacity] - [kube-reserved] - [system-reserved] - [hard-eviction-threshold] = 16000 - 0 - 1000 - 0 = 14900MB$

$Recommended\ Allocatable\ Resources = [Allocatable\ Resoures] * 0.75 = 14900 * 0.75$
$= \mathbf{11175\ MB\ RAM}$

- Therefore, three nodes are required to host this workload:

$$CPU = \frac{10000}{3500} \approx 3$$

$$Memory = \frac{32000}{11175} \approx 3$$

That means we will need **3 nodes with 4 vCPU and 16 GB RAM** to provide the capacity required for this application.

Storage sizing

Now, let's calculate the number of volumes required, as follows:

- **Virtual machines (VMs)**: 3 (nodes) * 120 GB (recommended per server) = **360 GB disk**

- Workload: **500 GB RWO**

- Internal registry: **200 GB RWX**

- Logging: **106 GB RWO (see next)**

  Frontend:

  $bytes\ per\ line * lines\ per\ second = 30 * 256 = 7680\ bytes\ per\ pod\ per\ second * 20\ pods$
  $= 153{,}600\ bytes\ per\ second = 153{,}600 * 60 * 60 * 24$
  $= 13{,}271{,}040{,}000\ bytes\ per\ day \approx 14\ GB\ per\ day * 7\ days$
  $= \textbf{98\ GB\ of\ logs\ in\ 7\ days}$

  Backend:

  $bytes\ per\ line * lines\ per\ second = 10 * 256 = 2560\ bytes\ per\ pod\ per\ second * 4\ pods$
  $= 10{,}240\ bytes\ per\ second = 10{,}240 * 60 * 60 * 24 = 884{,}736{,}000\ bytes\ per\ day$
  $\approx 1\ GB\ per\ day * 7\ days = \textbf{7\ GB\ of\ logs\ in\ 7\ days}$

  MongoDB:

  $bytes\ per\ line * lines\ per\ second = 1 * 256 = 256\ bytes\ per\ pod\ per\ second * 1\ pods$
  $= 256\ bytes\ per\ second = 256 * 60 * 60 * 24 = 22{,}118{,}400\ bytes\ per\ day$
  $\approx 0.03\ GB\ per\ day * 7\ days = \textbf{0.21\ GB\ of\ logs\ in\ 7\ days}$

  Total:

  $$Total = 98 + 7 + 0.21 \approx \textbf{106\ GB\ of\ logs\ in\ 7\ days}$$

- **Monitoring**: 248 GB RWO (as we saw in the previous section about the sizing for monitoring in a cluster up to 50 nodes and 1,800 Pods)

Summary

The following table summarizes the servers required for this cluster, considering three additional servers dedicated to hosting the OpenShift infrastructure components (**Logging, Monitoring, Registry, and Ingress**).

Server	Qty	vCPU	Total vCPU	RAM (GB)	Total RAM (GB)	DISK (GB)	Total DISK (GB)
Master	3	4	12	16	48	120	360
Infrastructure Worker (Internal Registry + Ingress + Monitoring + Logging)	3	8	24	32	96	120	360
Application Worker	3	4	12	16	48	120	360
Total	9		48		192		1080

In the previous table, the bootstrap node is not being considered as it is a temporary node that is removed after cluster installation.

And finally, the requirements for Persistent Volumes are summarized in the following table:

Tool	Volume Size (GB)	Type
Workload (MongoDB)	500	RWO
Logging	106	RWO
Internal Registry	200	RWX
Monitoring	248 (94 * 2 + 20 * 3)	RWO
Total RWO	854	RWO
Total RWX	200	RWX

Now that we already know some best practices to observe in an OpenShift cluster, let's discuss in the next section some surrounding aspects you should also consider when designing an OpenShift architecture.

Infrastructure/cloud provider

As the OpenShift platform is integrated with the infrastructure or cloud provider, some prerequisites are also required, but for now, during the architecture design phase, you basically need to define which provider you will go for and be aware that they have specific prerequisites. We are not covering these pre requisites in this chapter, as this is going to be explained in depth in *Chapter 5, OpenShift Deployment*.

In that chapter, we will practice the deployment process itself, starting by preparing the infrastructure or cloud prerequisites, setting up installer parameters, storage, network, the virtualization/cloud layer, and so on. However, during the architecture design phase, in general, you don't need to go deeper into these details yet, but just choose which provider to go for and keep in mind some specifications you will have to fulfill for the provider you have chosen.

Network considerations

An OpenShift cluster uses an SDN layer to allow communication between workloads and cluster objects. The default plugin used with OpenShift at the time this book was written is **OpenvSwitch (OvS)**, but OpenShift is also compatible (and supported) with the **OVN-Kubernetes** plugin. Check this link to better understand the differences between the plugins: `https://docs.openshift.com/ container-platform/latest/networking/openshift_sdn/about-openshift- sdn.html#nw-ovn-kubernetes-matrix_about-openshift-sdn`.

Within the SDN, there are two virtual subnets—the first one has the Internet Protocol (IP) addresses that a Pod inside the cluster uses, while the second is always used when you create a service object. The default values for these subnets are listed in the following table:

Network	Subnet ID	Mask bits
OpenShift Pods'Network	10.128.0.0	/14
OpenShift Services' Network	172.30.0.0	/16

> **Important Note**
>
> The preceding ranges are customizable during the platform installation process only! You cannot modify these after installation.
>
> Make sure these two ranges don't conflict with the existing one in your physical infrastructure. If you have conflicts, you may experience routing problems between Pods on OpenShift and external services that have a real IP within these ranges. The reason is simple: OpenShift SDN will always think that anything with an IP within the Pods' range is a pod inside the cluster—and in this case, the SDN will never deliver this package to the external network (network address translation, or NAT). Therefore, a pod on OpenShift will never be able to communicate with a real service out of the cluster that has an IP within the Pods' or services' range. So, be careful to define these two ranges with ones that will *never* be used in your infrastructure.

Let's move on to some other important aspects you need to consider from the network perspective, then!

VPC/VNet

If you are deploying OpenShift on **Amazon Web Services** (**AWS**), Azure, or **Google Cloud Platform** (**GCP**), you may choose to install an OpenShift cluster in a new or existing VPC/**virtual network** (**VNet**). If you go for existing VPC/VNet components such as subnets, NAT, internet gateways, route tables, and others, these will no longer be created automatically by the installer—you will need to configure them manually.

DNS

Depending on the installation method and the provider, different DNS requirements are needed. Again, we are going to cover this point in detail later in this book, but keep in mind that a set of DNS requirements depends on the provider and installation method you choose.

Load balancers

The *IPI* in on-premises environments already comes with an embedded highly available load balancer included. In cloud environments, OpenShift uses load balancers provided by the cloud provider (for example, AWS Elastic Load Balancing (ELB), Azure's Network LB, GCP's Cloud Load Balancing). With *UPI*, you need to provide an external load balancer and set it up before cluster deployment.

DHCP/IPMI/PXE

If you go for OpenShift on bare metal, observe other requirements specified for this type of environment. DHCP, IPMI, and PXE are optional; however, they are recommended to have a higher level of automation. Therefore, consider that in your cluster architectural design.

Internet access

The OpenShift platform needs download access from a list of websites—the Red Hat public registries to download the images used with it, either using a proxy or direct access. However, it is possible to install it on restricted networks as well. Additional work is required, though: you need to establish an internal registry first and mirror all required images from Red Hat's registries to there. If you use a proxy, also check the proxy's performance to avoid timeout errors during the image pulling process with OpenShift.

Well, we've covered great content so far, from foundation concepts to best practices you need to observe related to the installation mode, computing, network, and storage. We are almost done with the most important aspects of an OpenShift cluster architecture, but we can't miss some considerations related to authentication and security. See in the following section some final considerations we brought to this chapter to help you with your cluster's architecture design.

Other considerations

Finally, there are a few more things that you should also consider during the design phase of your OpenShift cluster.

SSL certificates

OpenShift uses SSL for all cluster communication. During the platform installation, self-signed certificates are generated; however, it is possible to replace the API and ingress certificates. At this point, you only need to know that this is possible; later in this book, you will see how to do it.

IdPs

OpenShift is deployed using a temporary `kubeadmin` user. It is highly recommended you configure new IdPs to allow users to log in to the platform using a convenient and safe authentication method. There are several supported IdPs with OpenShift; here is a current list of supported options (at the time of writing this book):

IdP	Description
HTPasswd	Uses `htpasswd` to configure a local authentication mechanism.
Keystone	To use OpenStack Keystone.
LDAP	To use standard LDAP IdP, using simple bind authentication.
Basic authentication	Basic authentication to validate credentials using a remote IdP. It is a sort of generic integration mechanism with an external provider.
Request header	Using request-header, such as `X-Remote-User`, to identify users. Usually used with an authentication proxy system, which sets the request header value.
GitHub or GitHub Enterprise	Validate usernames and passwords using GitHub as the authentication system.
GitLab	The same as with GitHub/GitHub Enterprise but using GitLab.
Google	Similar to GitHub or GitLab but using Google's OpenID Connect.
OpenID Connect	Integrate with any OpenID-compatible provider.

To wrap up this chapter and give you a quick reference guide, look at the OpenShift architectural checklists we provide next.

OpenShift architectural checklists

These checklists will help you define the main decisions you may need to take during the OpenShift architecture design and can also be used as a summary of the concepts covered in this chapter.

Here's a checklist for installation mode and computing:

Installation mode (select one only)	• IPI • UPI • Agnostic
Provider (select one only)	• AWS • Azure • GCP • VMware vSphere • OpenStack • Red Hat Virtualization • Bare metal • IBM Z • IBM Power • Other (agnostic). Specify: _____
Compute—Master Nodes	Number (define the number of masters you need for your cluster): CPU (define the amount of CPU for your master nodes): Memory (define the amount of CPU for your master nodes):
Compute—Infra Nodes	Number (define the number of infra nodes you need for your cluster): CPU (define the amount of CPU of your infra nodes): Memory (define the amount of CPU of your infra nodes):
Compute—Workers	Number (define the number of worker nodes you need for your cluster): CPU (define the amount of CPU for your worker nodes): Memory (define the amount of CPU for your worker nodes):

Here's a checklist of additional tools:

Aggregated logging tool (select one only)	• OpenShift Logging out of the box • Forward logs to an external logging tool
OpenShift monitoring (select one only)	• Using ephemeral volume • Using persistent volume
Service mesh (select one only)	• Not required to use/install • Required
Serverless (select one only)	• Not required to use/install • Required
Pipelines (Tekton) (select one only)	• Not required to use/install • Required
GitOps (ArgoCD) (select one only)	• Not required to use/install • Required

Here's a checklist for storage:

Storage provider— block (RWO) (multiple choices)	• Red Hat OpenShift Data Foundation (aka OpenShift Container Storage) • AWS Elastic Block Store (EBS) • Azure Disk • Cinder • Google Compute Engine (GCE) persistent disk • VMware vSphere • OpenStack Cinder • Internet Small Computer Systems Interface (iSCSI) • Fibre Channel (FC) • Local Volume • Hostpath • Other. Specify: _____
Storage provider— file (RWX) (multiple choices)	• Red Hat OpenShift Data Foundation (aka OpenShift Container Storage) • Network File System version 4 (NFSv4) • Azure Files • OpenStack Manila • Other. Specify: _____
Storage sizing	RWO (define the total size expected for RWO volumes considering the required amount for tools specified in the previous table): RWX (define the total size expected for RWX volumes):

Here's a checklist for the network:

VPC/VNet (select one only)	• New (provisioned by the installer) • Existing (configured manually)
Subnets	Pod (Define the subnet range for pods; default is 10.128.0.0/14): Service (Define the subnet range for services; default is 172.30.0.0/16): Machines (Define the subnet range for the nodes' IP; default is 10.0.0.0/16):
DNS records required:	API (Uniform Resource Locator (URL) that API will use and need to be configured on external DNS using A or Canonical Name (CNAME) record (for example, `api.<cluster_name>.<base_domain>.`): Ingress (default domain that applications will use and need to be configured on external DNS as a wildcard A or CNAME record (for example, `*.apps.<cluster_name>.<base_domain>.`) Others (if you have other custom domain that needs to be set up on external DNS):
External load balancer required? (select one only)	• N/A • Only for ingress • Both ingress and API
Other required network elements? (one or more)	• DHCP • IPMI • PXE • Others. Specify: _____
Environment connected? Using proxy? (select one only)	• Directly • Using proxy • Restricted

Here's a checklist for other general considerations:

Certificates: (select one for each)	API: • Self-signed • Custom Ingress: • Self-signed • Custom
IdP: (one or more)	• HTPasswd • Keystone • LDAP • Basic authentication • Request header • GitHub • GitLab • Google • OpenID Connect
Other integrations or considerations	(Use this space for other general considerations)

Summary

In this chapter, we went through some of the most important aspects you need to consider and define before starting a cluster deployment, at the architectural design phase. You now understand the different choices you have with the platform and how to estimate the number and size of your nodes and storage.

Check the next chapter—*Chapter 3, Multi-Tenant Considerations*—to acquire more knowledge about the multi-tenant aspects of the OpenShift architecture.

Further reading

If you want to go deeper into the topics we covered in this chapter, look at the following references:

- *etcd documentation:* `https://etcd.io/docs/latest/`

- *Kubernetes official documentation:* `https://kubernetes.io/docs/home/`

- *About Kubernetes Operators:* `https://kubernetes.io/docs/concepts/extend-kubernetes/operator/`

- *Documentation about* rpm-ostree: `https://coreos.github.io/rpm-ostree/`

- *CSI drivers supported by Open Container Platform (OCP):* `https://docs.openshift.com/container-platform/4.8/storage/container_storage_interface/persistent-storage-csi.html#csi-drivers-supported_persistent-storage-csi`

- *Graphical explanation about allocatable resources:* `https://learnk8s.io/allocatable-resources`

- *How to plan your environment according to application requirements:* `https://docs.openshift.com/container-platform/latest/scalability_and_performance/planning-your-environment-according-to-object-maximums.html#how-to-plan-according-to-application-requirements_object-limits`

- *Recommended host practices, sizing, and others:* `https://docs.openshift.com/container-platform/latest/scalability_and_performance/recommended-host-practices.html`

3
Multi-Tenant Considerations

As with almost any software, things get more difficult as you scale in number and size. In the previous chapter, we looked at the most important aspects related to the architecture of an OpenShift cluster. In this chapter, we are going to cover some things you should understand when working with multiple environments on one or more clusters.

In this chapter, we will cover the following topics:

- What is multitenancy?
- Handling multiple tenants
- Multitenancy on OpenShift
- Multi-tenant strategies

What is multitenancy?

Multitenancy is the ability to provide services for multiple users or groups (also known as **tenants**) using a single platform instance. A platform architecture can be designed to be single- or multi-tenant:

- In a single-tenant platform architecture, there is no isolation within an instance of the platform. This means that there is no separation in an instance and that, as such, there is no way to isolate objects for users or groups. In this case, to achieve multitenancy, you need to provision a separate platform instance for every tenant.

- In a multi-tenant platform architecture, it is possible to isolate objects and data between different tenants. Therefore, you can protect each tenant's data and objects, thus providing enough privacy and security, even by using a single platform instance.

Depending on the architectural design and how OpenShift is used, you can have both types of platforms (single- or multi-tenant). In this chapter, you will learn how to define the best way to consider this while designing your OpenShift clusters.

Handling multiple tenants

There are many different ways to work with multiple tenants on OpenShift, with the most obvious one being to have a single cluster for each tenant. However, this is not always possible or the best option: having dedicated hardware and a platform for every tenant can be costly, difficult to maintain, and not efficient. With shared clusters, multiple workloads from different tenants share the same computing capacity, enabling more efficient computing usage.

OpenShift can provide isolation for objects, computing, network, and other hardware resources for each tenant, ensuring they are isolated from each other. In the next section, we are going to look at the different types of isolation and how to utilize them.

Multitenancy in OpenShift

When it comes to multitenancy on OpenShift, several objects are involved. The following table shows some important resources that provide multi-tenant capabilities, all of which we are going to cover in this chapter:

Kubernetes resources	Limit Kubernetes resources between different applications, tenants, and users.	Namespaces and role-based access control (RBAC).
Computing	Limit computing resources for each tenant.	ResourceQuota, nodeSelector, Taint, and Tolerations.
Network	Limit network traffic between the different applications of each tenant.	NetworkPolicy and ingress sharding.
Storage	Limit which kind and amount of storage resources are allowed for each tenant.	ResourceQuota and StorageClass.

The following diagram illustrates how these resources are combined to provide isolation and enable multitenancy:

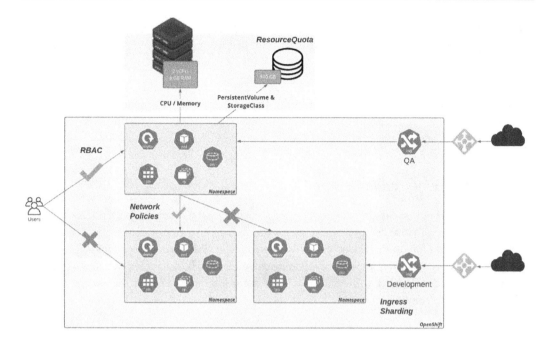

Figure 3.1 – Multitenancy and isolation

Now, let's look at how these objects are used to provide multitenancy capabilities on OpenShift. We are going to use all of them while covering practical examples from *Chapter 5, OpenShift Development*, onward, so don't worry about how to use them for now. Instead, focus on understanding how they provide isolation capabilities to enable multitenancy.

Namespaces

Namespaces provide some level of isolation. Using namespaces, you can define a limited space for the following:

- **Kubernetes workloads**: **Pods**, **Secrets**, **ConfigMaps**, **Deployments**, and so on.
- **Access control**: Isolate the namespace resource's access by giving the appropriate **roles** (permissions) to the users or groups.
- **Limit resource consumption**: It is possible to limit the number of resources that are consumed by a namespace using **ResourceQuotas**.

Role-based access control

Permission control over resources on OpenShift is done using **roles** and **RoleBindings**. **Roles** are a set of actions (such as `get`, `list`, `watch`, `create`, `upgrade`, and so on) that are permitted over resources (such as **Pods**, **Services**, **Deployments**, **Jobs**, and so on), while **RoleBindings** (or **ClusterRoleBinding**) are how you bind a role to a subject (groups, users, or **ServiceAccounts**).

You are required to use roles and RoleBindings to give users the right permissions to the right namespaces, according to the tenants and the separation logic you want to implement.

ResourceQuotas

ResourceQuotas allows a cluster administrator to constraint a namespace to a limited set of resources. It can limit computing resources and/or the number of objects. It is a crucial thing to consider when you're using shared clusters to ensure there's a limited capacity for each tenant. Without ResourceQuotas, only one namespace can consume the entire capacity of a worker node, for instance.

nodeSelectors, taints, and tolerations

Through **nodeSelectors**, you can dedicate workers for a specific reason or tenant. With nodeSelectors, it is possible to isolate physical compute resources for each tenant: in a 10-node cluster, you can have, for instance, five nodes for QA and the other five for development. **Taints** and **tolerations** are different ways of doing this: while with nodeSelectors, you instruct a Pod to be scheduled in a defined set of nodes that contain a certain label, with taints, you instruct a worker to *repeal* Pods that do not contain a certain toleration to run in it.

NetworkPolicy

A **NetworkPolicy** provides a standard way to isolate network traffic between Pods and namespaces. It works like a firewall in which you define ingress and/or egress policies to accept/deny traffic flows between different Pods and namespaces.

Ingress/router sharding

On OpenShift, you can create multiple ingress controllers, which will allow you to isolate ingress traffic between different tenants.

Multi-tenant strategies

It is important to understand that there is no physical isolation when using the multi-tenant objects listed previously – isolation is defined and implemented by the software. However, it is possible to provide a physical level of isolation by using different multi-tenant strategies, as you will see now. The best strategy depends on the requirements you have in your company; some companies care more about having an efficient use of computing resources, while others don't care about spending more resources in favor of the most secure isolation strategy possible.

Some different strategies are as follows:

- Dedicated clusters, one for each tenant
- A shared cluster with no physical separation of resources

- A shared cluster with dedicated worker nodes
- A shared cluster with dedicated worker nodes and ingress controllers

Dedicated clusters

As we have already mentioned, the most obvious strategy is to have a different cluster for each tenant. The following diagram shows an example of providing services for two tenants (QA and Development):

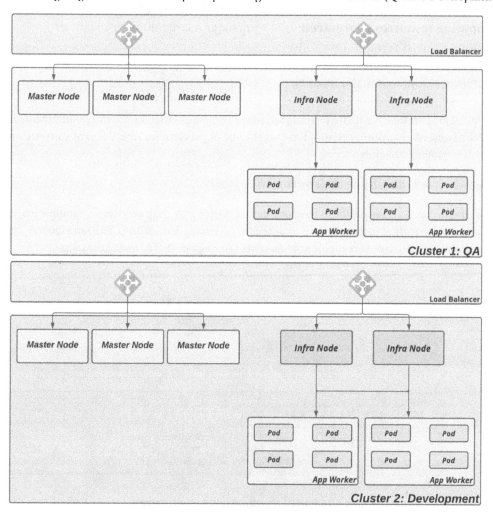

Figure 3.2 – Dedicated clusters

The following table contains a summary of the pros and cons of this strategy, from our point of view:

Pros	Cons
Highest isolation: Physical separation of the workloads, which ensures workloads from different tenants will not be able to communicate inside a cluster. **Computing resources guaranteed:** You don't need (if you don't want) to deal with ResourceQuotas to guarantee computing resources for the tenants.	**Higher resource usage:** For each cluster, you need at least three master nodes and likely some infra-nodes. **Operational overhead:** You may need to manage a large number of clusters.

This type of architecture is usually recommended for companies that have strict requirements for physically isolated environments and don't want to be dependent on multi-tenant software and processes to provide isolation.

Shared clusters, no physical separation

On the other hand, you may decide to have one shared cluster providing services for multiple tenants while using OpenShift objects to enable multitenancy (namespaces, RBAC, ResourceQuotas, and NetworkPolicies). You can see a simple schema of this strategy in the following diagram:

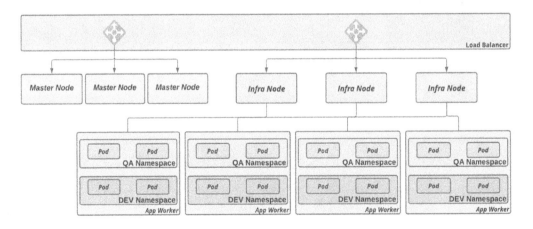

Figure 3.3 – Shared cluster, no physical separation

The following table shows some of the pros and cons of this strategy:

This kind of architecture usually works well for non-production environments, in which some incidents related to performance degradation, for instance, are sometimes tolerated.

Shared clusters, dedicated worker nodes

If you have to provide computing resources that are physically isolated, this may be the right direction to take. In this case, you are going to use the same objects to provide isolation, but you dedicate worker nodes for specific tenants. The following is a simple schema:

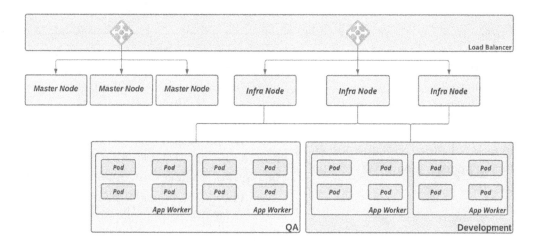

Figure 3.4 – Shared cluster, dedicated worker nodes

This strategy can have the following pros and cons:

Pros	Cons
Computing physical isolation: Dedicated worker nodes for specific tenants and, as such, have physical isolation of computing power.	**Less efficient resource usage:** You may have resources available in one tenant while another suffers from a lack of resources, for instance.
Operational effort: Save operational effort by managing only one (or a few) cluster(s).	**No ingress isolation:** Ingress for all tenants is provided by the same group of ingress instances, which means that a high traffic load in an application from a particular tenant may impact applications from other tenants.

This kind of architecture usually works well for non-production environments, in which some incidents related to performance degradation, for instance, are sometimes tolerated. This architecture also works well for production workloads that don't have requirements for high HTTP(S) throughput and low-latency HTTP(S). The following link provides a capacity baseline for a single OpenShift ingress (**HAProxy**), for comparison: `https://docs.openshift.com/container-platform/ latest/scalability_and_performance/routing-optimization.html`.

Pros	Cons
Lowest resource usage: As you don't need to have dedicated resources for each tenant, you may have more efficient use of computing resources. **Operational effort:** Save operational effort by managing only one (or a few) cluster(s).	**No physical isolation:** Isolation is done based on the namespace and other objects, which is highly dependent on the usage process. A wrong ResourceQuota setup, for instance, can lead to capacity problems; or a mistake in a NetworkPolicy configuration can result in inappropriate allowed/denied network traffic. These problems can be mitigated by automating the creation and management of Namespaces, ResourceQuotas, NetworkPolicies, and others using tools such as Ansible, Jira, Jenkins, ServiceNow, and more. **No ingress isolation:** Ingress for all the tenants is provided by the same group of ingress instances, which means that an application with a high traffic load from a particular tenant may impact applications from other tenants.

Shared clusters, dedicated worker nodes, and ingress controllers

Finally, with this strategy, you can share a cluster among different tenants by providing a higher level of isolation. The following diagram shows this cluster's architecture:

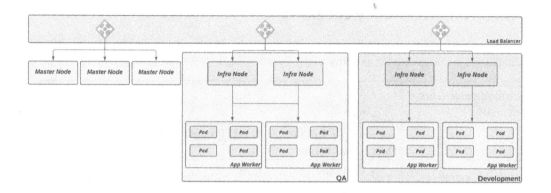

Figure 3.5 – Shared cluster, dedicated worker nodes, and ingress controllers

This strategy can bring the following pros and cons:

Pros	Cons
Computing physical isolation: Have dedicated worker nodes for specific tenants and, as such, have physical isolation of computing power. **Ingress isolation:** Each tenant has ingress instances, which ensure that a high traffic load in one tenant will not impact other tenants. **Operational effort:** Save operational effort by managing only one (or a few) cluster(s).	**Less efficient resource usage:** You may have resources that are available in one tenant while another suffers from a lack of resources. Additional infra-nodes may be required for the ingress of each tenant.

This kind of architecture usually works well for both production and non-production environments. If it's well defined and used, it has the same effect as dedicated clusters. The only difference between this strategy and dedicated clusters is that, in this case, a configuration mistake could lead to an application being deployed in the wrong tenant (by using the incorrect **nodeSelector** in a namespace, for instance).

OpenShift multitenancy checklist

To conclude this chapter, we have decided to add some additional items to the checklist we started building in the previous chapter, as follows:

Are multiple tenants required?	• Yes • No
Define the tenants.	For example, Development, QA, Production Dept A, Production Dept B, and so on.
Level of isolation required between tenants?	• Compute • Network • Ingress
Does the workload require high throughput and low latency for HTTP(s) requests?	• Yes • No
Decision.	• Dedicated clusters • Shared clusters using namespaces, RBAC, ResourceQuotas, and NetworkPolicies • Shared clusters with dedicated worker nodes • Shared clusters with dedicated worker nodes and ingress controllers
Strategy to create and maintain namespaces, RBAC, ResourceQuotas, and NetworkPolicies.	• Manual • Automated using _____ (specify)
Responsible for creating and maintaining namespaces, RBAC, ResourceQuotas, and NetworkPolicies.	Specify the team(s) that will be responsible for creating and maintaining the namespaces and other objects with the right labels, permissions, quotas, and so on. This ensures data privacy and security between different tenants.

In the next chapter, you will acquire more knowledge about the personas that are usually related to OpenShift, from C-level to operational level, and what skills are required for each role. You will be able to understand what you should expect from each role and prepare yourself and your team to work well with OpenShift.

Summary

In this chapter, we looked at some of the strategies that provide services for multiple tenants with OpenShift clusters. You now understand that we can have dedicated or shared OpenShift clusters to host tenants. You also saw that with shared clusters, you can provide some level of isolation for each tenant by using namespaces, ResourceQuotas, NetworkPolicies, and other objects to provide multitenancy or even have a physical separation of workers and/or ingress; the best option for your use case depends on the requirements of your organization, workloads, and environments.

However, I need to warn you that in the current hybrid cloud world, you will probably need to work with clusters in different providers and regions, which may lead you to have an increasing number of clusters. But don't worry – as we saw in *Chapter 1*, *Hybrid Cloud Journey and Strategies*, many great tools can help us manage several clusters, such as Red Hat Advanced Cluster Management, Advanced Cluster Security and Quay; we will take a deep dive into these tools by covering practical examples in the last part of this book, which is dedicated only to them.

In the next chapter, we will learn about the personas and skillsets that are usually related to OpenShift, their main duties and responsibilities, and other important factors.

Further reading

If you want to find out more about the concepts that were covered in this chapter, check out the following references:

- *What is multitenancy?* `https://www.redhat.com/en/topics/cloud-computing/what-is-multitenancy`

- *Multi-tenant network isolation (OpenShift documentation):* `https://docs.openshift.com/container-platform/4.8/networking/network_policy/multitenant-network-policy.html`

4
OpenShift Personas and Skillsets

Now that you understand what an OpenShift cluster architecture looks like, what about understanding the people that work with it? I have seen in my career many adoptions of different software, platforms, and processes, some of them running smoothly and others being challenging. What are the differences between them that define good or bad adoption? The main difference is the people – that is, how people are treated in the process is crucial to determining good adoption.

We strongly believe that you need to care about people as much as you care about the platform itself. Because of that, we decided to reserve some pages of this book to discuss the main personas and skills usually needed to perform activities well.

Therefore, you will find in this chapter the following:

- Some of the personas that are related to OpenShift
- The key responsibilities and duties that are usually required for those personas
- The challenges they usually face
- What they usually expect to see in a container platform
- A skills matrix containing the main abilities that a resource needs to have

The following are the main topics covered in this chapter:

- Personas
- The skills matrix

Let's dive in!

Personas

We are going to discuss in this section the professional roles (**personas**) that we have usually worked with in our career dealing with OpenShift and that, in general, are related to OpenShift on some level.

In the following figure, you will see some of the typical roles from an IT department:

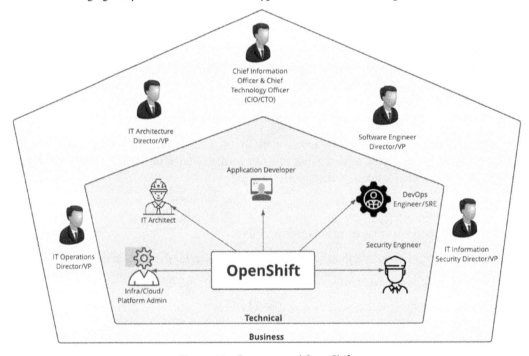

Figure 4.1 – Personas and OpenShift

Let's review some of the main technical roles and learn useful information that will allow us to better prepare an organization for OpenShift adoption. If you work in a coordinator role, take this opportunity to learn what you should expect from your technical team; if you are an individual contributor, learn what the activities and skills are that you will probably need to handle on a daily basis.

A system/cloud/platform administrator

A system, cloud, or platform administrator is an individual contributor who is responsible for maintaining the underlying infrastructure of the OpenShift clusters and also the clusters themselves. Depending on the scale of an organization, there are multiple specializations in this role, such as network, operating system, and storage administration.

Let's look at some of the duties they need to perform and the challenges they face while doing so.

Key responsibilities and duties

- Support the network, storage, and virtualization layers and the underlying infrastructure.
- OpenShift cluster administrative operations, such as scale nodes, upgrades, and granting permissions to users and groups.
- Configure OpenShift namespaces, network policies, resource quotas, node selectors, and so on.
- Platform capacity management.
- Cloud cost estimation and reporting.
- Open support tickets with vendors to get assistance to deal with issues related to the clusters' underlying infrastructure and/or OpenShift.
- Monitor the platform performance and stability.
- Analyze and report incidents proactively when noticing possible issues.
- Respond to users about incidents.
- Collect logs to investigate issues.
- Report metrics about operations, such as the number of incidents, **Mean Time to Recovery (MTTR)**, and **Mean Time Between Failures (MTBF)**.

The challenges they face (usually)

- **Standards**: As clusters grow, it becomes increasingly difficult to manage them mainly due to the lack of standards to use.

- **Observability**: Organizations that struggle to establish comprehensive and efficient application and infrastructure monitoring are not uncommon, unfortunately. Inefficient monitoring tools and practices make it much more difficult to make the right decisions. Administrators have difficulties in establishing tools and practices to observe environments properly.

- **Capacity planning and management**: Cluster administrators often have challenges with correctly managing the cluster's capacity. It is not an unusual situation where clusters run out of resources due to a lack of monitoring and capacity planning.

- **Cloud cost**: As cloud usage grows, keeping costs under control also becomes more difficult.

- **Supporting multiple cloud environments**: Administrators often need to deal with and support multiple cloud environments, which makes administration and management much more challenging.

- **Logging management**: Some companies struggle to define and maintain an efficient logging platform for containers, which impacts issue analysis, root-cause detection of a problem, and so on.

- **Support tickets**: It is common to have difficulties managing support tickets from vendors, sometimes due to lack of knowledge of how to collect logs and other things, such as insufficient information for a vendor to review, or vendor delay.

What do they want to see on a container platform?

- A mature enterprise container orchestrator platform to host the applications developed by the development teams.

- Vendors that can provide a great support experience and establish themselves as trusted advisor partners.

- A platform that is compatible with different infrastructure and/or cloud vendors.

- If there is a monitoring and/or logging tool in place, the platform needs to be able to integrate with it. If there isn't, the platform needs to have an out-of-the-box logging and monitoring tool.

- A platform with available GitOps tools, such as Argo CD.

- Monitoring tools and dashboards that can assist in the capacity planning and management of the clusters.

- A platform with a high level of automation out of the box and/or can easily be automated using tools such as Ansible or Terraform.

- A complete product documentation and knowledge base available.

- Great training that is available from vendors or partners.

The next persona that usually has a tight relationship with OpenShift is the IT architect. Let's take a look at it closely now.

IT architect

An IT architect, in the context of OpenShift, is an individual contributor who is responsible for designing the high-level architecture and definitions for one or more clusters. Architects collect business requirements and understand issues, problems, and other references to define a solution that best addresses them. There are several different architect roles, which change from company to company, but some usual ones are the following:

- **Application architects**: Design and supervise the development and release of applications and related technical tasks.

- **Solution architects**: In general, architects with strong technical but also management knowledge who lead entire projects.

- **Enterprise architects**: An architect who works with the business to define the whole IT strategy for the company.

- **Infrastructure/cloud architects**: Design and supervise projects on cloud providers or on-premises hardware and infrastructure.

The architect role described as follows is closer to an infrastructure/cloud architect, who generally will work much closer with OpenShift than the others.

Key responsibilities and duties

- Work with the business to get the requirements for a container orchestration platform.

- Plan and design the cluster architecture, considering the business requirements and best practices such as high availability, security, and other important aspects for any critical platform.

- Estimate sizes to handle the initial required capacity.

- Estimate the effort to deploy and support the platform.

- Estimate the initial and ongoing cost of the platform.

- Assist the director and others at the VP level on the platform-buying decision.

- Design a monitoring and logging solution for the platform.

- Define security requirements for the platform.

- Define the ongoing policies and process requirements, such as backup and disaster recovery.

- Assist the DevOps engineer to define **Continuous Integration/Continuous Deployment (CI/CD)** processes and pipelines using the container platform.

- Hand over the decisions and specifications to the development and operations teams. Assist in deployment and platform usage.

The challenges they face (usually)

- **Knowledge**: I have seen many times IT architects that were pushed to design architecture for an OpenShift cluster without knowing sometimes even the basic concepts about it. They are used to designing architectures for applications in other less complex systems and not for a platform such as Kubernetes or OpenShift, in which you need to deal with a wide range of aspects such as virtualization, storage, and network.

- **Lack of sponsorship**: Architects usually need to get information and engage people from different teams inside an organization to understand and define some aspects of the architecture – for example, which storage options they have available to use with the platform, which network characteristics will be used, and so on. When they don't have proper sponsorship to engage the right people, the architecture design usually is time-consuming and frustrating.

- **Requirement definitions**: It is not uncommon for organizations to not have a clear definition of what the requirements for the platform are. Some companies don't even know what they want to achieve with the platform and decide to adopt the technology just because some competitors are using it, or other similar reasons. That makes the architectural definition a much harder task, as it is difficult to be assertive and accurate with the decisions that need to be made in the process.

- **Vendor assistance**: Architects need a lot of assistance from the vendor during the platform's architectural design. Engaging with the vendor sometimes is challenging, due to a lack of availability, a delay in replying to information, and so on.

What do they want to see on a container platform?

- A mature enterprise container orchestrator platform to host the applications developed by the development teams
- Vendors that can meet service-level agreements for support
- Trusted advisor vendors that can assist with information, demonstrations, and proof of concepts
- A platform that is compatible with the infrastructure and/or cloud vendors defined by the business
- A platform that can integrate with existing tools, such as **Application Performance Management (APM)**, monitoring, and logging
- A comprehensive platform that includes monitoring, logging, CI/CD, service mesh, and serverless tools.
- GitOps tools available out of the box
- An ecosystem that includes IDE plugins, samples, and other development tools
- A platform with a high level of automation out of the box and/or can easily be automated using tools such as Ansible or Terraform
- A central management solution for multiple clusters
- A complete product documentation and knowledge base available
- Great training that is available from vendors or partners

A really important piece of this puzzle is the application developer; without them, there is no application to host on any platform! Let's take a look at the role now.

Application developer

A developer is an individual contributor who uses a container platform to build and deploy application code. In some companies, the application developer has direct access to the container platform to perform basic actions on it, such as deployment and configuration of the application in development and/or **Quality Assurance (QA)** environments; in other companies, they only have access to a CI/CD system to run a pipeline to check and deploy code automatically.

Key responsibilities and duties

- Understand functional requirements, use cases, or user stories to develop and test code.
- Develop code using best practices such as **Test-Driven Development** (**TDD**), CI, Infrastructure as Code, and GitOps.
- Commit code frequently.
- Develop and run unit tests.
- Build the container application.
- Support QA teams to test an application.
- Bug-fixing of production issues.

The challenges they face (usually)

- **Knowledge**: Usually, organizations don't give the appropriate concern to a developer's platform knowledge. They are well educated with code languages, frameworks, IDEs, and related subjects but not about the platform that will host the application – in this case, OpenShift. That is a frequent cause of rework in order to adapt the code to specific platform requirements.
- **Proper permissions**: It is not uncommon that developers don't have proper permission with a container platform to perform their duties. This impacts their performance and generates frustration.
- **Proper processes and tools**: Some developers don't have the proper tools and processes available to perform activities by themselves. Usually, they depend on other teams to get the proper IDE and runtimes installed, update the CI/CD pipeline to reflect a new change, and so on.
- **Logging management**: In some companies, developers are not allowed to have direct access to logs; they need to request it from another team, and sometimes, that takes a long time, impacting bug-fixing issues.

What do they want to see in a container platform?

- Plugins for the most popular IDE
- A desktop development environment with tools to build and test containerized applications
- Comprehensive samples with a wide range of software development stacks

- Container base images with different supported runtimes (such as Java, Spring Boot, Node.js, and Quarkus)

- Seamless integration with the CI/CD tool

- The UI developer console, in which they can see the application logs, monitoring, and configurations

- A complete product documentation and knowledge base available

- Great training that is available from vendors or partners

- A platform with proper permissions set that allows them to perform their duties

Some more recent roles are also closely related with OpenShift such as the DevOps engineer and the **Site Reliability Engineer** (SRE). In the following section, you will learn more about them.

The DevOps engineer/SRE

A DevOps engineer is usually the individual contributor responsible for developing and maintaining CI/CD pipelines to support development teams. They also analyze possible issues that occur with application build and deployments, using the CI/CD and container platform tools. This role can bring a lot of discussions, as DevOps itself is a culture and a set of practices to bring infrastructure and development together to be more agile and efficient, rather than a job role.

Similarly, an SRE is usually the one that brings some software engineering aspects to infrastructure to automate repetitive operational tasks and enables scalable and reliable environments.

DevOps and SRE are tightly related – an SRE can be represented as one specific implementation of DevOps practices, usually focusing more on infrastructure and operational problems and implementations, while the DevOps engineer usually works more with the CI/CD process, supporting development teams to deliver code more efficiently.

We have decided to conflate both roles into one persona but adapt it to your real-world environment, and consider the following information to see which best fits your role in your organization.

Key responsibilities and duties

- Understand CI/CD processes and develop pipelines accordingly.
- The continuous enhancement of CI/CD pipelines that support the development teams.
- Define, build, and deploy standards for containers along with CI/CD pipelines.
- Deploy and support tools used with CI/CD processes, such as Container Registry and the Git versioning tool.
- Automate processes related to CI/CD (such as the DNS and load balancer configurations).
- Support development teams regarding questions and issues concerning CI/CD pipelines.
- Automate repetitive operational tasks such as server provisioning and network configurations.
- Define and implement observability of services.
- Capacity planning.
- Plan and run chaos testing.

The challenges they face (usually)

- **Knowledge**: DevOps engineers and SREs are also usually well educated with automation-related things but not always with the container platform and its underlying infrastructure.
- **Proper processes and tools**: The best tools to develop a desired CI/CD, testing, and automation are not always available. In these situations, they are usually pushed to use workarounds, develop things with much more effort, and so on, which can be really frustrating!
- **Platform limitations**: These professionals often need to automate processes with legacy systems that don't have APIs, which makes automation development much more difficult.

What do they want to see in a container platform?

- Easy integration with existing CI/CD tools
- A UI console in which they can see the build and deployment logs and check the deployment objects
- A high level of automation out of the box
- A CLI and API that helps to automate manual steps
- Plugins for CI/CD tools that facilitate pipeline development

- An easy way to deploy an application over multiple clusters

- Out-of-the-box observability or one that can be easily plugged into an existing tool

- A platform that helps to implement perturbation models for chaos testing

Finally, we have the security engineer, who is always important in any process, even more so now in the age of ransomware and other real security concerns. Let's check the following section to see how they are related to OpenShift.

The security engineers

The security engineer is an individual contributor responsible for reviewing processes and tools that provide security for systems and data. They are always evaluating applications and systems to control environments from known threats, protect existing infrastructure to minimize vulnerabilities, and finally, detect and respond against attackers.

Key responsibilities and duties

- Evaluate possible platform vulnerabilities and create strategies to mitigate them.

- Review CI/CD pipelines and add security gates to maximize security.

- Establish a container image security scanning tool to detect any known vulnerability within images.

- Define security policies to be used with all clusters.

- Ensure that clusters are compliant with defined policies.

The challenges they face (usually)

- **Knowledge**: The security engineer is no different from the previous roles. They usually have real concerns with OpenShift, mostly because they are not well educated with it and, as such, don't understand how security is implemented in it.

- **Secrets management**: Management of passwords and sensitive data is still a challenge on Kubernetes. There are several vault systems available in the market already, but many companies still don't have a well-established practice and the tools to support it, which makes this a point of concern for security teams.

- **Permissions management**: It is a real challenge to define the right permissions level to ensure a secure platform that doesn't lock down developers and other teams, impacting their productivity.

- **Vulnerability scan**: Security tools for containers and Kubernetes are fairly recent, so most companies don't have any available, which can cause real nightmares for security engineers!

What do they want to see in a container platform?

- A platform that uses the most secure standards for container orchestration, such as Linux namespaces, `cgroups`, Linux capabilities, `seccomp`, and SELinux

- A tool that allows them to easily define and enforce security policies among several clusters

- A multi-tenant secure-capable platform that enables network isolation between different workloads

- Data encryption at rest

- A secure **Role-Based Access Control (RBAC)** platform available

- An audit log enabled and logging

- Image registry with a security scan

We have covered the different technical roles and their responsibilities, which will help you and your organization to be more prepared for OpenShift adoption. Finally, we will present a skills matrix that aims to help understand what the main skills required by each role are.

The skills matrix

In this section, you will find a skills matrix suggested for the roles listed in the preceding sections. The items listed here are a subset of technical skills that we have observed from experience as being important for a professional to know. Note that we are not covering soft skills in this book; this doesn't mean that we don't place any importance on soft skills – on the contrary, we do! – however, they are not the focus of this book. The skills matrix is classified in the following criteria:

■ (Basic)	Basic knowledge – able to perform basic tasks independently
■■ (Advanced)	Advanced level – has the skills to complete more complex tasks with little or no assistance
■ / ■■ (Expert)	Expert level – has the skills to complete complex tasks and also pass knowledge on
■■ / ■■ (Specialist)	Specialist level – expert level but also able to work on optimizations and solution design for complex problems around a topic, and seen as a subject matter expert by other colleagues and partners

Check out the skills in the following tables, which we have split into parts for better readability.

Architecture-, infrastructure-, and automation-related skills

Check the following table for the skills needed for architecture, infrastructure, and automation tasks:

Skill	Platform administrator	IT architect	Application developer	DevOps engineer/SRE	Security engineer
Incident and problem handling	Expert	Advanced	Basic	Advanced	Advanced
System and software architecture design/modeling	N/A	Specialist	N/A	Basic	Advanced
Cloud service providers (AWS, Azure, GCP, and so on)	Advanced	Expert	Advanced	Expert	Expert
Networks, operating systems, and storage	Specialist	Advanced	Basic	Advanced	Expert
Automation tools (Ansible, Terraform, and so on)	Specialist	Basic	Basic	Expert	Advanced
Capacity planning	Advanced	Expert	Basic	Basic	Basic

Development-, container-, and CI/CD-related skills

In the following table, you will find the skills related to development and CI/CD:

Skill					
Container building tools (Docker Compose build, S2I, Buildah, and BuildConfig)					
Container runtime (Docker, Podman, and so on)					
CI/CD tools (Jenkins, GitLab CI, Tekton, and so on)					
Infrastructure as code/GitOps (Argo CD, Flux CD, and so on)					
Logging tools (ELK, EFK, Loki, and so on)					
Monitoring/APM tools (Prometheus, Grafana, Zabbix, AppDynamics, Dynatrace, Amazon CloudWatch, Azure Monitor, and so on)					
Container image registries (Quay, Harbor, AWS ECR, Azure CR, Docker Hub, Artifactory, Nexus, and so on)					
Programming languages used by the organization (Java, Python, and so on)					
Databases used by the organization (MySQL, MongoDB, and so on)					

OpenShift-related skills

Finally, see the following table for the skills required for OpenShift:

OpenShift installation (*)			N/A	N/A	
Authentication and authorization					
Container Network Interface for OpenShift/ K8s (OVS/OVN/Ingress/Egress)					
Storage for OpenShift/K8s (in-tree/CSI)					
OpenShift Day 2 (scaling nodes, upgrades, operators' deployment, and so on)					

(*) If administrators have assistance from a consulting partner to deploy the clusters. If they are required to deploy new clusters, then they should have at least an advanced level of OpenShift installation.

Summary

We have seen in this chapter the main personas related to OpenShift, their key responsibilities, the challenges they usually face, and what they look for in a container platform. Now, we also have an idea of the competencies an OpenShift professional needs to have. We expect it to help you prepare yourself for the tasks you will perform as an individual contributor or plan your team enablement and hiring policy accordingly, if you are at a management level.

This chapter concludes the first part of this book, reserved for concepts, architecture, and enablement. Now, go to the next chapter and start your journey on OpenShift cluster deployment. In that chapter, we will walk you through the steps to prepare the installation, deploy a cluster, and perform some post-deployment tasks.

Further reading

If you want to look at more information related to the concepts we covered in this chapter, check the following references:

- *What is a skills matrix and should I use one in 2021?*: `https://www.skills-base.com/what-is-a-skills-matrix-and-should-i-use-one-in-2021`

- *Everything-as-Code*: `https://openpracticelibrary.com/practice/everything-as-code/`

- *GitOps*: `https://openpracticelibrary.com/practice/gitops/`

- *How Full is My Cluster? Capacity Management and Monitoring on OpenShift*: `https://cloud.redhat.com/blog/full-cluster-capacity-management-monitoring-openshift`

- *A layered approach to container and Kubernetes security*: `https://www.redhat.com/en/resources/layered-approach-container-kubernetes-security-whitepaper`

- *OpenShift Container Platform architecture*: `https://docs.openshift.com/container-platform/4.9/architecture/architecture.html`

Part 2 –
Leverage Enterprise Products
with Red Hat OpenShift

In this part, you will be introduced to the OpenShift deployment and Day 2. You will be driven to understand how to build an OpenShift cluster and explore the objects and APIs that will help you to perform administrative tasks. You will be guided through different scenarios with important insights to protect the health of a cluster, grant adequate access policies, and avoid common mistakes that often cause overcommitting and so compromise a cluster.

This part of the book comprises the following chapters:

- *Chapter 5, OpenShift Deployment*
- *Chapter 6, OpenShift Troubleshooting, Performance, and Best Practices*
- *Chapter 7, OpenShift Network*
- *Chapter 8, OpenShift Security*

5

OpenShift Deployment

In this chapter, we will navigate through a deployment process for Red Hat OpenShift. A successful implementation case will be only possible when you understand the needs of the architecture, understand the branding technology applied to the underlying infrastructure, and understand the workloads you are supposed to have on top of OpenShift.

As you have seen in *Chapter 1, Hybrid Cloud Journey and Strategies*, there are a lot of options in this multi-cluster universe. So, what are the best options for you? How can you choose? It can be very confusing if you start to prepare the requisites of the deployment without properly preparing yourself for it. That said, we must stress the importance of being aligned with the architecture chosen to ensure the expected deployment will succeed. Runtime changes and unplanned gaps can have unexpected results and create a platform full of flaws, causing the platform to malfunction. Making a comparison with the real world, it is like a ship that cannot be smoothly loaded with containers because it is always being repaired and, as such, is not ready to perform long trips or even depart.

Considering the explanation in the first part of this book, there are now several options for deploying OpenShift clusters. Next, we'll start with a checklist of questions that will help you make the right decision for your needs. Keep in mind, there are no right or wrong answers, but it will help you decide which is the best for you.

This chapter covers the following:

- Requirements
- OpenShift installation prerequisites
- Preparing for the installation
- Installation
- What's next?

Let's get started then!

Requirements

This is a practical chapter in which you will deploy an OpenShift cluster using what we consider to be the most complex deployment procedure: the **User-Provisioned Installer** (**UPI**)/agnostic installation.

The source code used in this chapter is available at `https://github.com/PacktPublishing/ OpenShift-Multi-Cluster-Management-Handbook/tree/main/chapter05`.

As we already covered in this book, there are many different types of installations and supported providers, and it is almost impossible to cover every combination of them – neither is it our intention, as there is plenty of documentation, tutorials, and great references on the internet that will guide you through all types of installations.

That said, we understand that the most-added-value deployment procedure we can bring to you is the UPI/agnostic one; when you make it, you will be able to understand and easily execute the other types of installations. The reason for this is simple: with the UPI/agnostic installation, you are responsible for providing all the prerequisites that an OpenShift deployment requires, while with the **Installer-Provisioned Infrastructure** (**IPI**) deployment, the installer itself will provide the prerequisites for you automatically within the underlying provider.

OpenShift checklist opt-in

Have you seen the checklist we gave to you in *Chapter 2*, *Architecture Overview and Definitions*, and *Chapter 3*, *Multi-Tenant Considerations*. If you are reading this chapter to implement an OpenShift cluster in an enterprise and you haven't read those chapters yet, we strongly recommend you go back and read them, as they contain important aspects you need to think about and consider first, before deploying a cluster.

Lab requisites

To follow the labs in this chapter, you will need the following:

- A hypervisor or cloud provider in which you can spin up the instances. You can also use bare metal servers if you have them available.

- This is the minimum requirement for the VMs:

  - One temporary server for the Bootstrap node

  - Three servers for master nodes, with 2 vCPU, 8 GB RAM, and 50 GB of disk (minimum)

  - Two servers for worker nodes, with 2 vCPU, 8 GB RAM, and 50 GB of disk (minimum)

  - One server for the bastion node, with 2 vCPU, 4 GB RAM, and 20 GB of disk (minimum)

If you don't have enough resources available in your environment, you can also use a three-node cluster, in which masters and workers are co-located in the same nodes.

> **Important Note**
>
> The requirements listed are valid only for a lab. Refer to *Chapter 2, Architecture Overview and Definitions*, to get a reliable sizing for an enterprise installation.

OpenShift installation prerequisites

Before starting your journey through OpenShift deployment, you must observe several prerequisites. First, we will explore the options according to the decision that you previously made in the opt-in form.

As explained in *Chapter 1, Hybrid Cloud Journey and Strategies*, OpenShift has three installation modes: **Installer-Provisioned Infrastructure (IPI)**, **User-Provisioned Infrastructure (UPI)**, and **agnostic** (that is, a bare metal installer). It is very important to remember that no option will work well for every case, although the best option is the one that best fits into the architecture designed for you previously.

This chapter is very focused on deployment and all things related to spawning up a cluster by yourself, so keep that in mind when making your own lab and enjoy the tips and materials on our GitHub repository, which will be a real Swiss Army knife for you.

The following table shows you which installation methods you have available, according to the provider chosen (at the time this book was written):

Provider	IPI	UPI	Agnostic
VMware ESXi 6.7+	Recommended	Possible	N/A
Nutanix AHV	Recommended	Not Available	N/A
Red Hat OpenStack	Recommended	Possible	N/A
Red Hat Virtualization	Recommended	Possible	N/A
Amazon Web Services (AWS)	Recommended	Possible	N/A
Microsoft Azure	Recommended	Possible	N/A
Microsoft Azure Stack Hub	Recommended	Possible	N/A
Google Cloud Platform (GCP)	Recommended	Possible	N/A
IBM Z	Not Available	Recommended	N/A
IBM Power	Not Available	Recommended	N/A
Bare Metal	Recommended	Not Available	Possible

Regarding the terms in the previous table, we classified some of the available options for each infrastructure provider to give you an overview of the current possibilities (at the time of writing this book). When we say **Recommended**, we are not only giving our perspective, but we are trying to say this is a *common and best choice* for that scenario. **Possible** indicates a valid option, but you will have some penalties, such as losing some great automation features the product brings out of the box. For that reason, we classified them as **Possible**, but not as the best choice. **Not Available** is self-explanatory.

As the prerequisites will be different according to the installation method, we prepared a matrix that helps you start the preparation of the underlying infrastructure to begin the cluster deployment:

Installation Method	Prerequisites	Common Requisites
UPI Installation	<ul><li>DHCP</li><li>DNS</li><li>Web Server</li><li>Load Balancer</li><li>Firewall Rules</li></ul>	<ul><li>Pull Secret</li><li>OpenShift Installer</li><li>OpenShift Client</li><li>SSH Key</li></ul>
Agnostic Installation	<ul><li>All UPI prerequisites + PXE Server (optional, you can boot from ISO instead)</li></ul>	
IPI Installation	<ul><li>Provider properly configured (Account permissions and limits)</li></ul>	

UPI/agnostic installer

Any OpenShift installation must have a valid Red Hat subscription, the OpenShift installer binaries, a **pull secret** file, a public **Secure Shell (SSH) key**, and the resources available according to each provider.

In this section, we will guide you through a feasible and reliable cluster installation, whatever provider you have chosen. We will also set up the prerequisites needed using practical examples – feel free to use those configurations as many times as you need.

Note that the files used in this chapter are also available in our GitHub repository: `https://github.com/PacktPublishing/OpenShift-Multi-Cluster-Management-Handbook/tree/main/chapter05`

So, let's start with the prerequisite systems that are not part of OpenShift itself, but are indispensable to ensure that everything will work fine during the deployment process.

DNS

In this section, we will discuss the **Domain Name System** (**DNS**) requirements to provision an OpenShift cluster. For demonstrations purposes, we will give the minimum configuration to make everything work; for in-depth settings, check the references we have provided in the last chapter of this book.

For our lab, we will use the **BIND** tool running in a Red Hat Enterprise Linux 8 VM; however, you can use any other DNS server on top of Windows or your preferred Linux distribution. We will refer to this Linux VM from now on by the name **bastion**, which is a kind of convention with Red Hat architectures. If you want to strictly follow the instructions in this chapter, we recommend you use a fresh installation of Red Hat Enterprise Linux 8, using the minimum install.

An OpenShift cluster requires a dedicated subdomain. To facilitate your understanding, we will use a hypothetical *hybrid cloud company* that uses hybridmycloud.com as its main public domain. The complete subdomain for the OpenShift cluster will be ocp.hybridmycloud.com.

To install BIND, run the following commands on your bastion VM:

```
$ sudo yum install bind bind-utils -y
$ sudo systemctl enable --now named
$ sudo firewall-cmd --permanent --add-port=53/tcp
$ sudo firewall-cmd --permanent --add-port=53/udp
$ sudo firewall-cmd --reload
```

Now, we are going to configure the DNS server to be used with the OpenShift installation and applications. Perform the following steps to accomplish this:

1. Create a subdomain zone by adding the following code in your named.conf file. You can alternatively download a file ready to be used in our GitHub repository at chapter05/named.conf:

    ```
    $ sudo cat << EOF >>  /etc/named.conf
    zone "ocp.hybridmycloud.com" IN {
    type master;
    file "/var/named/ocp.hybridmycloud.com.db";
    allow-query { any; };
    allow-transfer { none; };
    allow-update { none; };
    };

    zone "1.168.192.in-addr.arpa" IN {
    type master;
    ```

```
file "/var/named/1.168.192.in-addr.arpa";
allow-update { none; };
};
EOF
```

2. Create a forward zone file at /var/named/ocp.hybridmycloud.com.db:

```
$ sudo cat <<EOF > /var/named/ocp.hybridmycloud.com.db
;[1] Begin Common Header Definition
\$TTL 86400
@ IN SOA bastion.ocp.hybridmycloud.com. root.ocp.
hybridmycloud.com. (
202201010001 ;Serial
21600 ;Refresh
3600 ;Retry
604800 ;Expire
86400 ;Minimum TTL
)
;End Common Header Definition

;Name Server Information [2]
    IN NS bastion.ocp.hybridmycloud.com.

;IP address of Name Server [3]
bastion IN A 192.168.1.200

;api internal and external purposes [4]
api        IN    A    192.168.1.200
api-int    IN    A    192.168.1.200

;wildcard application [5]
*.apps     IN    A    192.168.1.200

;bootstrap node to start cluster install only [6]
bootstrap  IN    A    192.168.1.90

;master nodes [7]
```

```
master1      IN    A     192.168.1.91
master2      IN    A     192.168.1.92
master3      IN    A     192.168.1.93

;worker nodes [8]
worker1      IN    A     192.168.1.101
worker2      IN    A     192.168.1.102
EOF
```

Let's look at this code in more detail:

[1]: Common DNS zone header.

[2]: The nameserver will be its own Bastion server.

[3]: The IP address from the nameserver (Bastion IP).

[4]: These records are mandatory and need to point to the VIP that will be used for the OpenShift API functions. In our case, we are using the bastion server as the VIP (suitable only for lab environments).

[5]: Wildcard VIP record used for the applications that run on OpenShift. In our case, we are using the bastion server as the VIP (suitable only for lab environments).

[6]: Bootstrap node IP record, used only for the cluster installation and can be removed after it.

[7]: Master node IP records, where the control plane objects will be hosted.

[8]: Worker node IP records, where the workloads will run. If you go for a three-node cluster, disregard the worker hosts.

3. Create a reverse zone file at /var/named/1.168.192.in-addr.arpa:

```
$ sudo cat <<EOF > /var/named/1.168.192.in-addr.arpa
\$TTL 1W @    IN     SOA     bastion.ocp.hybridmycloud.com.
root (
2019070700 ; serial
3H             ; refresh (3 hours)
30M            ; retry (30 minutes)
2W             ; expiry (2 weeks)
1W )           ; minimum (1 week)

5.1.168.192.in-addr.arpa. IN PTR
api.ocp.hybridmycloud.com.;
5.1.168.192.in-addr.arpa. IN PTR
api-int.ocp.hybridmycloud.com.;
```

```
90.1.168.192.in-addr.arpa. IN PTR
bootstrap.ocp.hybridmycloud.com.;
91.1.168.192.in-addr.arpa. IN PTR
master1.ocp.hybridmycloud.com.;
92.1.168.192.in-addr.arpa. IN PTR
master2.ocp.hybridmycloud.com.;
93.1.168.192.in-addr.arpa. IN PTR
master3.ocp.hybridmycloud.com.;
101.1.168.192.in-addr.arpa. IN PTR
worker1.ocp. hybridmycloud.com.;
102.1.168.192.in-addr.arpa. IN PTR
worker2.ocp. hybridmycloud.com.;
EOF
```

> **Important Notes**
>
> Do *not* create a reverse zone record for the application's wildcard VIP, as that will lead to the wrong DNS resolution.
>
> If you created it for a three-node cluster, disregard the worker A and PTR records.

4. Restart the named service:

    ```
    sudo systemctl restart named
    ```

5. Validate the DNS to ensure that all DNS records are set up appropriately using the following `dig` commands (replace `192.168.1.200` with your bastion IP):

    I.   Validate the OpenShift API using the following:

    ```
    dig +short @192.168.1.200 api.ocp.hybridmycloud.com
    dig +short @192.168.1.200 api-int.ocp.hybridmycloud.com
    ```

    II. For the BIND samples we described in this section, the output *must* be as follows:

    ```
    192.168.1.5
    192.168.1.5
    ```

    III. Validate the application's wildcard using the following:

    ```
    dig +short @192.168.1.200 joedoe.apps.ocp.hybridmycloud.
    com
    dig +short @192.168.1.200 whatever.apps.ocp.
    hybridmycloud.com
    ```

IV.   All results *must* point to the ingress application's wildcard VIP, as follows:

```
192.168.1.6
192.168.1.6
```

V.   Validate the nodes, as follows:

```
dig +short @192.168.1.200 boostrap.ocp.hybridmycloud.com
dig +short @192.168.1.200 master1.ocp.hybridmycloud.com
dig +short @192.168.1.200 master2.ocp.hybridmycloud.com
dig +short @192.168.1.200 master3.ocp.hybridmycloud.com
dig +short @192.168.1.200 worker1.ocp.hybridmycloud.com
dig +short @192.168.1.200 worker2.ocp.hybridmycloud.com
```

The answer must be the following:

```
192.168.1.90
192.168.1.91
192.168.1.92
192.168.1.93
192.168.1.101
192.168.1.102
```

VI.   Finally, let's validate the reverse records, as follows:

```
dig +short @192.168.1.200 -x 192.168.1.90
dig +short @192.168.1.200 -x 192.168.1.91
dig +short @192.168.1.200 -x 192.168.1.92
dig +short @192.168.1.200 -x 192.168.1.93
dig +short @192.168.1.200 -x 192.168.1.101
dig +short @192.168.1.200 -x 192.168.1.102
```

The results look similar to the following:

```
bootstrap.ocp.hybridmycloud.com.
master1.ocp.hybridmycloud.com.
master2.ocp.hybridmycloud.com.
master3.ocp.hybridmycloud.com.
worker1.ocp.hybridmycloud.com.
worker2.ocp.hybridmycloud.com.
```

Well done! If your DNS server is properly resolving names, you took a big step in preparing the prerequisites. Now, let's move on to another important piece of an OpenShift installation using the UPI method: the **Dynamic Host Configuration Protocol (DHCP)**.

DHCP

DHCP is used to provide IP addresses to the OpenShift nodes. In UPI or agnostic installation nodes, the IP address needs to be set using static configuration on DHCP (the `fixed-address` parameter).

Make sure that the IP address and hostname for the nodes in the DNS and DHCP match – each IP address and hostname in the DNS and DHCP need to be the same. In this subsection of prerequisites, we are focusing on creating a simple DHCP setup for later study and laboratory use. As previously stated, DHCP will be configured to provide static IP addresses, under the `192.168.1.x` subnet, so, this configuration uses the **media access control (MAC)** address of each node's Ethernet interfaces:

1. Install DHCP on your bastion VM:

    ```
    $ sudo yum install dhcp-server -y
    ```

2. Configure the `dhcpd.conf` file according to the hostnames and IP addresses used with the DNS:

    ```
    cat <<EOF > /etc/dhcp/dhcpd.conf
    # DHCP Server Configuration file.
    #[1]
    ddns-update-style interim;
    ignore client-updates;
    authoritative;
    allow booting;
    allow bootp;
    allow unknown-clients;
    default-lease-time 3600;
    default-lease-time 900;
    max-lease-time 7200;
    #[2]
    subnet 192.168.1.0 netmask 255.255.255.0 {
    option routers 192.168.1.254;
    option domain-name-servers 192.168.1.200;
    option ntp-servers 192.168.1.200;
    next-server 192.168.1.200; #[2.1]
    #filename "pxelinux.0";#[2.2]
    #[3]
    ```

```
group {
host bootstrap {
hardware ethernet 50:6b:8d:aa:aa:aa;
fixed-address 192.168.1.90;
option host-name "bootstrap.ocp.hybridmycloud.com";
allow booting;
}
host master1 {
hardware ethernet 50:6b:8d:bb:bb:bb;
fixed-address 192.168.1.91;
option host-name "master1.ocp.hybridmycloud.com";
allow booting;
}
host master2 {
hardware ethernet 50:6b:8d:cc:cc:cc;
fixed-address 192.168.1.92 ;
option host-name "master2.ocp.hybridmycloud.com";
allow booting;
}
host master3 {
hardware ethernet 50:6b:8d:dd:dd:dd;
fixed-address 192.168.1.93 ;
option host-name "master3.ocp.hybridmycloud.com";
allow booting;
}
host worker1 {
hardware ethernet 50:6b:8d:11:11:11;
fixed-address 192.168.1.101;
option host-name "worker1.ocp.hybridmycloud.com";
allow booting;
}
host worker2 {
hardware ethernet 50:6b:8d:22:22:22;
fixed-address 192.168.1.102;
option host-name "worker2.ocp.hybridmycloud.com";
allow booting;
```

```
        }
        }
        }
        EOF
        $ sudo systemctl enable --now dhcpd
        $ sudo firewall-cmd --add-service=dhcp --permanent
        $ sudo firewall-cmd --reload
```

Let's look at this code in more detail:

[1]: Common settings to define DHCP as authoritative in that subnet and times of IP lease.

[2]: Scope subnet definition:

- [2.1] and [2.2]: Must be defined when using a PXE server, helpful for bare metal installations. In this lab, we are going to use VMs and, as such, that will not be used; therefore, leave it commented (using the # character at the beginning of the line).

[3]: A group with all nodes to lease IP addresses. If you go for a three-node cluster, disregard the worker hosts.

Important Note

After you create the VMs in your hypervisor, update dhcpd.conf accordingly with the MAC addresses you get from the network interfaces; otherwise, no IP address will be given to this subnet.

Web servers

A web server is used to serve the OS image to install nodes, and also to provide the Ignition files (Ignition files are manifest files encoded on base64). In our scenario, we will install and configure an Apache web server, which is a very simple way to provide all the necessary tools for the cluster installation to run adequately.

Follow this short list of steps to accomplish this task:

1. Install an **httpd** server:

    ```
    $ sudo yum install httpd policycoreutils-python-utils -y
    ```

2. Configure /etc/httpd/conf/httpd.conf to change the **Listen directive**:

    ```
    $ sudo sed -i 's/80/81/g' /etc/httpd/conf/httpd.conf
    ```

3. Apply **SELinux** to change the default `httpd` port:

    ```
    $ sudo semanage port -a -t http_port_t -p tcp 81
    ```

4. Create a rule on **firewalld** to allow port `81`:

    ```
    $ sudo firewall-cmd --add-port 81/tcp --permanent
    $ sudo firewall-cmd --reload
    ```

5. Create a directory for OS image and Ignition files, and a file to test the connectivity:

    ```
    $ sudo mkdir -p /var/www/html/images
    $ sudo mkdir -p /var/www/html/ignition
    $ sudo touch /var/www/html/images/imageFileToTest.txt
    $ sudo touch /var/www/html/ignition/ignitionFileToTest.txt
    ```

6. Set permission and owner to the files:

    ```
    $ sudo chown -R apache. /var/www/html/
    $ sudo chmod 744 -R /var/www/html/
    ```

7. Start and enable the Apache web server:

    ```
    $ sudo systemctl enable --now httpd
    ```

8. Test connectivity using the `curl` command:

    ```
    $ curl -O http://192.168.1.200:81/images/imageFileToTest.
    txt
    $ curl -O http://192.168.1.200:81/ignition/
    ignitionFileToTest.txt
    ```

If you can see the file downloaded to your current folder, it means you have the Apache web server properly configured to serve the OpenShift installation.

Load balancer

A load balancer is another important element in an OpenShift cluster architecture. It is responsible for balancing the connection to the pool member in a set of nodes. For performance and resilience reasons, it is recommended that you use dedicated load balancing appliance hardware for production environments.

For our lab, we will be using HAProxy in our bastion VM to perform the load balancing function, which is a powerful, lightweight, and easy-to-use software load balancer.

That being said, before we start to configure it, it is important that you understand the basics of load balancing methods and the best practices that fit with the OpenShift platform.

I suppose the load balancer is like drops of water evenly distributed between a few cups, and each drop must fall into one glass, then the next drop must land in the next glass, and so on. Nonetheless, the cups are periodically dumped to avoid waste or overload.

So, there are some ways to perform the task; these ways are known as **balancing methods**. The following table explains the scenarios that OpenShift will make use of:

Method	Overview	OSI Model
Round Robin	Distribute the requests among all available servers on a cyclical basis. It is commonly the default setup for load balancers.	Layer 4 (Transport)
Least Connection	New connection to the least member in the pool or members with the least number of active connections.	Layer 4 (Transport)
Fastest	Least number of current outstanding sessions.	Layer 4 (Transport) and Layer 7 (Application)

A typical load balancer configuration is composed of four pairs of frontend and backend configurations that will balance the different types of requests and give a reliable fault tolerance to the platform.

The first pool members are the master nodes; these are a group of three members and should work with the Least Connection method with a sourced address setup. This setting ensures that, during internal API calls to the load balancer, the request will be handled from the same node that started the call requisition and, as such, gives the proper callback and asynchronous functioning.

You can find the HAProxy frontend and backend configurations in the sample here: (frontend `openshift-api-server` and backend `openshift-api-server`).

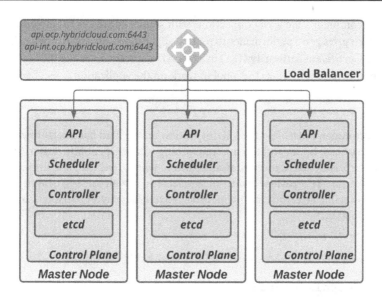

Figure 5.1 – Master node load balancer

The second group of load balancer configurations is also used with the master nodes for the OpenShift `machine-config-server` API. See the HAProxy configuration for the `machine-config-server` API in the frontend `machine-config-server` and backend `machine-config-server`.

The third and fourth groups of load balancers should be a pool of at least two nodes (worker nodes) that will be responsible for the traffic routing from the outside to the application distributed on the worker nodes of the cluster (one for HTTP and another for HTTPS).

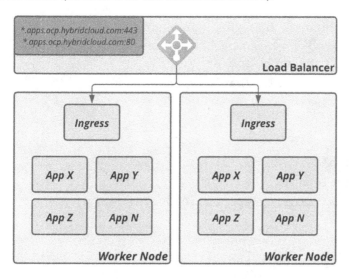

Figure 5.2 – Ingress load balancer

OpenShift often recommends the least connections with source addresses running on the transport layer (Layer 4,) which gives good performance to routing applications. However, when using certificates signed by a public **Certificate Authority (CA)** in the load balancer, instead of the OpenShift Ingress Controller, you must eventually set up this pool to work on the application layer (Layer 7).

> **Important Note**
>
> We strongly recommend disabling the SSL inspection on the load balancer/firewall layers to avoid digital certificate issues and the malfunctioning of the OpenShift cluster. This occurs because the enterprise load balancer/firewall solution, when used with SSL inspection mode enabled, decrypts every TCP payload and re-encapsulates them with a new SSL header. OpenShift interprets it as a certificate error, causing incorrect source/destiny TCP packets, and hanging up to TLS termination.

In a nutshell, the complete `haproxy.cfg` will be similar to the following:

```
$ sudo yum install haproxy -y
$ sudo cat <<EOF > /etc/haproxy/haproxy.cfg
# Global settings
global
  maxconn 20000
  log /dev/log local0 info
  chroot /var/lib/haproxy
  pidfile /var/run/haproxy.pid
  user haproxy
  group haproxy
  daemon
  # turn on stats unix socket
  stats socket /var/lib/haproxy/stats
defaults
  mode http
  log global
  option httplog
  option dontlognull
  option forwardfor except 127.0.0.0/8
  option redispatch
  retries 3
  timeout http-request 10s
  timeout queue 1m
```

```
  timeout connect 10s
  timeout client 300s
  timeout server 300s
  timeout http-keep-alive 10s
  timeout check 10s
  maxconn 20000
# Enable haproxy status endpoint
listen stats
  bind :9000
  mode http
  stats enable
  stats uri /
# OpenShift API (port 6443)
frontend openshift-api-server
  bind *:6443
  default_backend openshift-api-server
  mode tcp
  option tcplog
backend openshift-api-server
  balance source
  mode tcp
# bootstrap line below can be removed after the cluster is
deployed
  server bootstrap 192.168.1.90:6443 check
  server master1 192.168.1.91:6443 check
  server master2 192.168.1.92:6443 check
  server master3 192.168.1.93:6443 check
# machine-config-server API (port 22623)
frontend machine-config-server
  bind *:22623
  default_backend machine-config-server
  mode tcp
  option tcplog
backend machine-config-server
  balance source
  mode tcp
```

```
# bootstrap line below can be removed after the cluster is
deployed
    server bootstrap 192.168.1.90:22623 check
    server master1 192.168.1.91:22623 check
    server master2 192.168.1.92:22623 check
    server master3 192.168.1.93:22623 check
# Applications HTTP (port 80)
frontend ingress-http
    bind *:80
    default_backend ingress-http
    mode tcp
    option tcplog
backend ingress-http
    balance source
    mode tcp
    server worker1 192.168.1.101:80 check # [1]
    server worker2 192.168.1.102:80 check # [1]
# Applications HTTPS (port 443)
frontend ingress-https
    bind *:443
    default_backend ingress-https
    mode tcp
    option tcplog
backend ingress-https
    balance source
    mode tcp
    server worker0 192.168.1.101:443 check # [1]
    server worker1 192.168.1.102:443 check # [1]
EOF
```

Let's look at this code in more detail:

[1]: If you go for a three-node cluster, you should also point to the master nodes here.

After that, apply the HAProxy configuration by starting and enabling the service, as follows:

```
$ sudo setsebool -P haproxy_connect_any=1
$ sudo systemctl enable --now haproxy
$ sudo firewall-cmd --add-service=http --permanent
$ sudo firewall-cmd --add-service=https --permanent
$ sudo firewall-cmd --add-port=6443/tcp --permanent
$ sudo firewall-cmd --add-port=22623/tcp --permanent
$ sudo firewall-cmd --reload
```

After it starts, your load balancer can receive incoming connections and give the redirect to one member of the pool.

> **Note**
>
> As soon as the installation of the OpenShift control plane is finished, you will receive a message saying that it is safe to remove Bootstrap from load balancing; then, you can open the haproxy. cfg file and comment, or remove the lines that are referencing the Bootstrap server and restart the HAProxy server to apply the configuration.

Firewall

As the cluster has a lot of components involved, it is important to think about the security between Red Hat OpenShift and all other systems it integrates with. Unless you are working with a disconnected environment (refer to the *Further reading* section of this chapter for more details), you will need to grant access to ports 80 and 443 from certain URLs. These URLs are needed to download the required container images and others. Therefore, whitelist the following URLs in your network firewall:

URL	Required for
`registry.redhat.io`	Core container images.
`quay.io`	Core container images.
`*.quay.io`	Core container images.
`sso.redhat.com`	Red Hat SSO authentication URL.
`*.openshiftapps.com`	**Red Hat Enterprise Linux CoreOS (RHCOS)** images.
`cert-api.access.redhat.com`	Cluster Telemetry and Insights.
`api.access.redhat.com`	Cluster Telemetry and Insights.
`infogw.api.openshift.com`	Cluster Telemetry and Insights.
`console.redhat.com/api/ingress`	Cluster Telemetry and Insights.
`mirror.openshift.com`	Access mirrored installation content and images and release image signatures.
`storage.googleapis.com/openshift-release`	Source of release image signatures.
`*.apps.<cluster_name>.<base_domain>`	Default cluster routes to an ingress wildcard. It depends on the cluster and domain name chosen; for the purpose of this book, we are using `*.apps.ocp.hybridmycloud.com`.
`quay-registry.s3.amazonaws.com`	Quay image content in AWS.
`api.openshift.com`	Cluster token and to check if updates are available for the cluster.
`art-rhcos-ci.s3.amazonaws.com`	Repository to download Red Hat Enterprise Linux CoreOS (RHCOS) images.
`console.redhat.com/openshift`	Cluster token.
`registry.access.redhat.com`	Used by the odo CLI tool.

Reference

Refer to this link for the latest set of URLs: `https://docs.openshift.com/container-platform/4.9/installing/install_config/configuring-firewall.html`.

PXE server

A PXE server is a component that easily allows the boot process to look for installation files. During PXE configuration, you can create a simple **Grand Unified Bootloader** (**GRUB**) menu that works exactly like an OS installer, with all the kernel parameters you need.

We will deploy some packages to install the PXE server, create directories to store the configuration files, and develop a simple and useful bootstart menu. Now, SSH to your bastion server and do the following:

1. Install these `syslinux` packages:

    ```
    $ sudo yum install -y syslinux-tftpboot syslinux-nonlinux
    syslinux tftp-server
    $ firewall-cmd --add-service=tftp --permanent
    $ firewall-cmd --reload
    $ sudo systemctl enable --now tftp
    ```

2. Create the following directories:

    ```
    $ mkdir -p /var/lib/tftpboot/networkboot/coreOS
    $ mkdir -p /var/lib/tftpboot/pxelinux.cfg
    ```

3. Copy the required PXE server files, as follows:

    ```
    $ cp /usr/share/syslinux/* /var/lib/tftpboot
    ```

4. Copy the Red Hat CoreOS image files, as shown here (files are available for download at this link: `https://console.redhat.com/openshift/install/platform-agnostic/user-provisioned`):

    ```
    ├── networkboot
    │    └── coreOS
    │            ├── rhcos-live-initramfs.x86_64.img
    │            ├── rhcos-live-kernel-x86_64
    │            └── rhcos-live-rootfs.x86_64.img
    ```

5. Finally, create the bootloader menu to assist you in the installation:

    ```
    $ sudo cat <<EOF > /var/lib/tftpboot/pxelinux.cfg/default
    UI vesamenu.c32
    MENU COLOR sel 4 #ffffff std
    MENU COLOR title 0 #ffffff
    TIMEOUT 120
    ```

```
PROMPT 0

MENU TITLE OPENSHIFT 4.X AGNOSTIC PXE MENU

LABEL BOOTSTRAP NODE
  KERNEL networkboot/coreOS/rhcos-live-kernel-x86_64
  APPEND initrd=networkboot/coreOS/rhcos-live-initramfs.
x86_64.img,networkboot/coreOS/rhcos-live-rootfs.x86_64.
img coreos.inst.install_dev=/dev/sda coreos.inst.
ignition_url=http://192.168.1.200:81/ignition/bootstrap.
ign

LABEL MASTER NODE
  KERNEL networkboot/coreOS/rhcos-live-kernel-x86_64
  APPEND initrd=networkboot/coreOS/rhcos-live-initramfs.
x86_64.img,networkboot/coreOS/rhcos-live-rootfs.x86_64.
img coreos.inst.install_dev=/dev/sdacoreos.inst.ignition_
url=http://192.168.1.200:81/ignition/master.ign

LABEL  WORKER NODE
  KERNEL networkboot/coreOS/rhcos-live-kernel-x86_64
  APPEND initrd=networkboot/coreOS/rhcos-live-initramfs.
x86_64.img,networkboot/coreOS/rhcos-live-rootfs.x86_64.
img coreos.inst.install_dev=/dev/sdacoreos.inst.ignition_
url=http://192.168.1.200:81/ignition/worker.ign
EOF
```

Now that we have all the prerequisite components correctly set, we can start the installation using the UPI or agnostic installation method! So, go ahead and start your engines!

IPI

Even though we tried to give you a smooth demonstration of how to create all the prerequisite systems and servers, it might still look like an exhausting process. It is important to emphasize that it can be tiring when preparing on your own, but, in large enterprises, those infrastructures are often already working and need only some small setting tweaks to reach the necessary state.

You must be worn out after going through all the previous steps for the UPI installer. The good news is that the IPI installer is much easier to follow! You probably compared all the things needed on the table **Preparation Stuff Table**. To accomplish the task using the IPI, you should have only your cloud credentials, ensure the object limit of your cloud provider is enough for the minimum required in

OpenShift, choose the size of cloud instances that best fit your needs, create the `install-config.yaml` file, and run the OpenShift install binary to spawn your cluster.

This process is simple due to the high level of automation that OpenShift has under the hood, which uses cloud APIs to create all the prerequisites for you, according to the parameters you set in `install-config.yaml`. Obviously, there are some changes from cloud to cloud. In the following code, we show two excerpts that change in `install-config.yaml` when preparing your file for AWS, Azure, and GCP:

Here's an AWS `install-config` sample file:

```
apiVersion: v1
baseDomain: hybridmycloud.com
credentialsMode: Mint
controlPlane:
  hyperthreading: Enabled
  name: master
  platform:
    aws:
      zones:
      - us-west-2a
      - us-west-2b
      rootVolume:
        iops: 4000
        size: 500
        type: io1
        type: m5.xlarge
    replicas: 3
compute:
  - hyperthreading: Enabled
  name: worker
  platform:
    aws:
      rootVolume:
        iops: 2000
        size: 500
        type: io1
      type: c5.4xlarge
      zones:
```

```
        - us-west-2c
      replicas: 3
metadata:
  name: test-cluster
  networking:
    clusterNetwork:
    - cidr: 10.128.0.0/14
    hostPrefix: 23
    machineNetwork:
    - cidr: 10.0.0.0/16
    networkType: OpenShiftSDN
    serviceNetwork:
    - 172.30.0.0/16
  platform:
    aws:
      region: us-west-2
      userTags:
      adminContact: jdoe
    costCenter: 7536
    amiID: ami-96c6f8f7
    serviceEndpoints:
      - name: ec2
        url: https://vpce-id.ec2.us-west-2.vpce.amazonaws.com
fips: false
sshKey: ssh-ed25519 AAAA...
pullSecret: '{"auths": ...}'
```

Next, let's look at a sample Azure `install-config` file:

```
apiVersion: v1
baseDomain: hybridmycloud.com
controlPlane:
  hyperthreading: Enabled
  name: master
  platform:
    azure:
      osDisk:
```

```
          diskSizeGB: 1024
          type: Standard_D8s_v3
          replicas: 3
      compute:
      - hyperthreading: Enabled
      name: worker
      platform:
        azure:
          type: Standard_D2s_v3
          osDisk: diskSizeGB: 512
          zones:
          - "1"
          - "2"
          - "3"
      replicas: 5
metadata:
  name: test-cluster
  networking:
  clusterNetwork:
  - cidr: 10.128.0.0/14
  hostPrefix: 23
  machineNetwork:
  - cidr: 10.0.0.0/16
  networkType: OpenShiftSDN
  serviceNetwork:
  - 172.30.0.0/16
platform:
  azure:
    BaseDomainResourceGroupName: resource_group
    region: centralus
    resourceGroupName: existing_resource_group
    outboundType: Loadbalancer
    cloudName: AzurePublicCloud
pullSecret: '{"auths": ...}'
```

Here, we have a GCP `install-config` sample:

```
apiVersion: v1
baseDomain: hybridmycloud.com
controlPlane:
  hyperthreading: Enabled
  name: master
  platform:
  gcp:
    type: n2-standard-4
    zones:
    - us-central1-a
    - us-central1-c
  osDisk:
    diskType: pd-ssd
    diskSizeGB: 1024
    encryptionKey:
      kmsKey:
      name: worker-key
      keyRing: test-machine-keys
      location: global
      projectID: project-id
  replicas: 3
compute:
- hyperthreading: Enabled
  name: worker
  platform:
  gcp:
    type: n2-standard-4
    zones:
    - us-central1-a
    - us-central1-c
    osDisk:
    diskType: pd-standard
    diskSizeGB: 128
    encryptionKey:
    kmsKey:
```

```
      name: worker-key
      keyRing: test-machine-keys
      location: global
      projectID: project-id
    replicas: 3
metadata:
name: test-cluster
networking:
  clusterNetwork:
  - cidr: 10.128.0.0/14
    hostPrefix: 23
  machineNetwork:
  - cidr: 10.0.0.0/16
  networkType: OpenShiftSDN
  serviceNetwork:
  - 172.30.0.0/16
platform:
gcp:
  projectID: openshift-production
  region: us-central1
pullSecret: '{"auths": ...}'
fips: false
sshKey: ssh-ed25519 AAAA...
```

Well done! Now you have the correct `install-config.yaml` files to use with your cloud provider. Continue with the installation to start OpenShift using your preferred installation method.

Preparing for the installation

As you have seen in the previous sections, the prerequisites are very important, and any mistake could be an *Achilles' heel* for the OpenShift cluster installation and functioning. Failure to prepare the prerequisites correctly will cause errors during the cluster deployment that are not always easy to troubleshoot to find the root cause. That said, we would like to stress the importance of preparing and validating the pre-requisites correctly before starting the cluster deployment.

To start the installation using the UPI method, you will need the following:

- An SSH key pair
- A pull secret for the cluster, which you can generate by accessing `https://console.redhat.com/openshift/install`, with a valid user subscription
- OpenShift installer binary
- OpenShift command-line tools
- Installation configuration file (`install-config.yaml`)

In the following sections, we will detail all of those steps.

An SSH key pair

Starting from OpenShift version 4, Red Hat begins to use Red Hat CoreOS as the main OS due to the container and immutable nature. Red Hat CoreOS needs some Ignition files to provision the OS based on that configuration. This process leads to a secure and reliable way of provisioning OpenShift nodes, allowing a standard **zero-touch provisioning** (**ZTP**) process.

SSH is used to access the nodes directly and only through a pair of keys assigned to the username `coreos` (it is not possible to access the nodes using a simple username/password combination). It is vital to keep a copy of the SSH key pair used during the cluster deployment in case of a problem with your cluster, as this is the only way to directly access the nodes to collect logs and try to troubleshoot them. Also, the SSH key pair will become part of the Ignition files and the public key pair is distributed across all nodes of the cluster.

We are going to use our bastion VM to create an SSH key pair, by using the following command:

```
$ ssh-keygen -t ecdsa -N '' -f ~/.ssh/clusterOCP_key
```

We will use a public key in the next steps, for example, `clusterOCP_key.pub`.

> **Important Note**
> Never expose or share the SSH private key; any malicious person with the private key could get root access to the nodes and, with some knowledge, escalate privileges as an OpenShift `cluster-admin` user.

Pull secret

A pull secret is a file that contains a collection of usernames and passwords encoded in `Base64` used for authentication in image registries, such as `quay.io` and `registry.redhat.io`. You need to have a valid username at `console.redhat.com` to download or copy the pull secret.

To do so, complete the following two steps:

1. Access https://console.redhat.com/openshift/create and access **Downloads** in the side menu, as shown in the following figure:

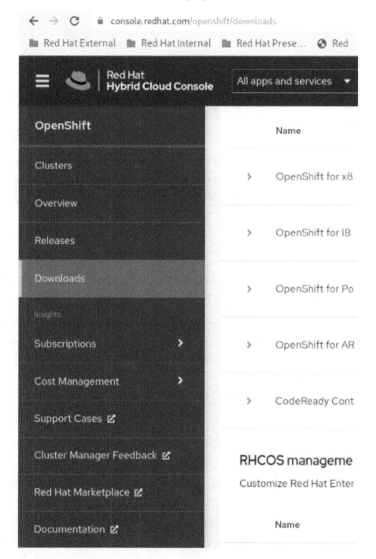

Figure 5.3 – Pull secret, Downloads menu

2. Scroll down to the **Tokens** section and click on the **Copy** or **Download** buttons to get the pull secret, as shown here:

Tokens

Figure 5.4 – Download or copy the pull secret

On this page, you will also find the command line, developer tools, and the installer binaries to download.

OpenShift installer binary

To install the cluster, you should download the installer binary, which can be downloaded from the Red Hat Hybrid Cloud Console, as shown in *Figure 5.3*, or you can browse the public repository found here: `https://mirror.openshift.com/pub/openshift-v4/x86_64/clients/ocp/latest/`.

OpenShift command-line tools

Like the installer binary, you can download the command-line tools under the same public repository mentioned previously, or through the Red Hat Hybrid Cloud Console from where you downloaded the pull secret.

OpenShift command-line tools consist of the `oc` and `kubectl` CLIs, which you will use to manage and run commands on the cluster as soon as it has been spawned.

Installation configuration file (install-config.yaml)

The last step before starting the installation is to create a manifest file called `install-config.yaml`. Essentially, this file consists of the control plane, workers, and network definitions, along with some metadata, such as the pull secret and the public SSH key.

Based on our previous configuration, the following is a sample of the `install-config` file that can be used with the installation. You can find this file in our GitHub repository at `chapter05/none-install-config.yaml`:

```
$ mkdir ~/ocp
$ cat <<EOF > ~/ocp/install-config.yaml
apiVersion: v1
baseDomain: hybridmycloud.com # [1]
compute:
- hyperthreading: Enabled
```

```
   name: worker
   replicas: 2 # [2]
controlPlane:
   hyperthreading: Enabled
   name: master
   replicas: 3 # [3]
metadata:
   name: ocp # [4]
networking:
   clusterNetwork:
   - cidr: 10.148.0.0/14 # [5]
     hostPrefix: 23
   networkType: OpenShiftSDN # [6]
   serviceNetwork:
   - 10.153.0.0/16 # [7]
platform:
   none: {} # [8]
fips: false
pullSecret: '<YOUR-PULL-SECRET>' # [9]
sshKey: '<YOUR-SSH-KEY>' # [10]
EOF
```

Let's look at this code in more detail:

[1]: The base domain for your cluster. Needs to match with the one you configured in your DNS in the previous section.

[2]: The initial number of worker nodes you will be deploying with the cluster. If you go for a three-node cluster, this value must be zero.

[3]: The initial number of master nodes you will be deploying with the cluster. For a highly available cluster, always use three replicas.

[4]: The cluster name. This also needs to match with what you specified in your DNS.

[5]: A block of IP addresses that will be used internally for pods within OpenShift SDN. We explored this concept in *Chapter 2, Architecture Overview and Definitions*.

[6]: The type of SDN used. Valid values are OpenShiftSDN or OVNKubernetes.

[7]: A block of IP addresses that will be used internally for network services within OpenShift SDN. We explored this concept in *Chapter 2, Architecture Overview and Definitions*.

[8]: Specific data about the underlying platform. This will be different depending on the platform on which your cluster will be hosted (such as AWS, Azure, GCP, and VMware). For agnostic installation, use none.

> **Note**
>
> You can use also the openshift-installer binary to generate a sample install-config.yaml file for the provider you are going to work with. Use the following command for that: ./openshift-install create install-config.

After this file is created, you can proceed to the installation steps in the following section.

Installation

Now, some further steps should be performed before deploying the cluster. The first step in the installation is to create the manifest files. We strongly recommend you create a backup of the install-config.yaml file before running the following command, as this command removes the original file and you will need to create it again from scratch if you need to retry the installation:

```
$ ./openshift-install create manifests --dir=home/user/ocp/
```

Open the ~/ocp/manifests/cluster-scheduler-02-config.yml file in your preferred editor. Change the mastersSchedulable parameter to false if you go for a regular cluster, or true if you decided to provision a three-node cluster.

You should now generate the Ignition files by running the following command:

```
$ ./openshift-install create ignition-configs --dir=home/user/ocp/
```

After the previous command, you should have three new Ignition files: bootstrap.ign, master.ign, and worker.ign.

Copy these three files to the HTTP server you prepared in the previous section:

```
$ sudo cp ~/ocp/*.ign /var/www/html/ignition
$ sudo chown -R apache. /var/www/html
$ sudo chmod 744 /var/www/html
```

Finally, you can proceed with provisioning the nodes.

Phase 1 – Provisioning servers

First, you need to provision the servers. This will vary depending on the underlying infrastructure but, in general, the process for virtualized environments (for example, VMware, vSphere, and RHV) is as follows:

1. Import the Red Hat CoreOS template to the hypervisor.

2. Clone it and configure the VM parameters according to the provider.

On the other hand, the process for bare metal or agnostic installation is performed either by booting using the Red Hat CoreOS ISO or using a PXE.

In our lab, we are going to boot using the Red Hat CoreOS ISO. Follow these steps to do it:

1. Download the ISO file from `console.redhat.com`, as mentioned previously, or directly through this link: `https://mirror.openshift.com/pub/openshift-v4/dependencies/rhcos/latest/rhcos-live.x86_64.iso`.

2. In the bastion VM, run the following command to get the `SHA512` digest of the Ignition files (this will be used after booting from the ISO to validate the authenticity of the file):

    ```
    $ sha512sum ~/ocp/bootstrap.ign
    $ sha512sum ~/ocp/master.ign
    $ sha512sum ~/ocp/worker.ign
    ```

    An example of the output is shown here:

    ```
    a5a2d43879223273c9b60af66b44202a1d1248fc01cf-
    156c46d4a79f552b6bad47bc8cc78ddf0116e80c59d2ea9e32ba53b-
    c807afbca581aa059311def2c3e3b
    installation_directory/bootstrap.ign
    ```

3. Boot using the ISO image, but don't specify any options until you see a shell prompt.

4. Run the following `coreos-installer` command to start the ignition process and, consequently, the OS provisioning:

    ```
    $ sudo coreos-installer install --ignition-url=
    http://192.168.1.200:81/ignition/bootstrap.ign /dev/sda
    --ignition-hash=SHA512-
    a5a2d43879223273c9b60af66b44202a1d1248fc01cf-
    156c46d4a79f552b6bad47bc8cc78ddf0116e80c59d2ea9e32ba53b-
    c807afbca581aa059311def2c3e3b
    ```

5.   Repeat the same process for each server, always respecting the following format for the `coreos-intaller` command:

```
$ sudo coreos-installer install --ignition-url=http://
192.168.1.200:81/<node_type>.ign <device> --ignition-
hash=SHA512-<digest>
```

Where `<node_type>` will be `bootstrap.ign`, `master.ign`, or `worker.ign`, `<device>` is the disk to be used to install the OS (such as `/dev/sda`), and `<digest>` is the result of the `sha512sum` command mentioned previously.

After booting the bootstrap and master nodes using this procedure, you can go to the next step to monitor the progress of the installation.

Phase 2 – Bootstrap and control plane

In this phase, Bootstrap will download the container images to provision the control plane components. As soon as the containers in each master node are running, the control plane components will start configuring themselves until the **etcd** cluster, API, and controllers from OpenShift are synchronized.

Run the following command from the Bastion VM to monitor the progress of the Bootstrap and control plane deployment:

```
./openshift-install wait-for bootstrap-complete --dir= /home/
user/ocp/ --log-level=debug
```

Immediately after the command is triggered, you will see some log messages on the console, similar to this sample:

```
INFO Waiting up to 30m0s for the Kubernetes API at https://api.
ocp.hybridmycloud.com:6443... INFO API v1.22.1 up INFO Waiting
up to 30m0s for bootstrapping to complete... INFO It is now
safe to remove the bootstrap resources
```

After that, you must remove Bootstrap from the load balancer and restart the `haproxy` service.

Note

Remember, the Bootstrap server is a one-shot use only; therefore, you can destroy the Bootstrap server completely from the infrastructure provider because it will not be used anymore, even if something goes wrong during the cluster installation.

Phase 3 – Check for certificates to sign – For UPI and agnostic installations only

When Bootstrap finishes its process, the Red Hat OpenShift Container Platform creates a series of **certificate signing requests (CSRs)** for each of the nodes. During our planning, we attempted to provision two worker nodes, so we must accept the certificates to join the worker nodes to the cluster.

We need to use the `oc` client to approve the certificates. To do so, run the following command to export the `kubeadmin` credentials and get access to the cluster:

```
$ export KUBECONFIG=~/ocp/auth/kubeconfig
```

A simple command can list the pending certificates and approve them until no pending certificates are showing:

```
$ oc get csr | grep -i Pending
NAME        AGE    REQUESTOR                                    CONDITION
csr-bfd72 5m26s system:node:worker0.ocp.hybridmycloud.com
Pending
csr-c57lv 5m26s system:node:worker1.ocp.hybridmycloud.com
Pending
. . .
```

Then, to approve the certificates, run the following command:

```
$ oc get csr -o name | xargs oc adm certificate approve
certificatesigningrequest.certificates.k8s.io/csr-bfd72
approved
certificatesigningrequest.certificates.k8s.io/csr-c57lv
approved
```

To confirm that everything worked fine, run the following commands until all nodes remain Ready:

```
$ oc get nodes
NAME                        STATUS    ROLES    AGE    VERSION
ocp-7m9wx-master-0          Ready     master   77d
v1.21.1+9807387
ocp-7m9wx-master-1          Ready     master   77d
v1.21.1+9807387
ocp-7m9wx-master-2          Ready     master   77d
v1.21.1+9807387
ocp-7m9wx-worker-jds5s      Ready     worker   77d
v1.21.1+9807387
```

```
ocp-7m9wx-worker-kfr4d    Ready    worker    77d
v1.21.1+9807387
```

Phase 4 – Finishing the installation

We are almost at the end of our UPI/agnostic installation! Now, we must check the cluster operators to ensure that all of them are available.

Using the following command, you will be able to monitor the cluster operators' deployment progress:

```
./openshift-install wait-for install-complete --dir= /home/
user/ocp/ --log-level=debug
INFO Waiting up to 30m0s for the cluster to initialize...
```

When it finishes, you will receive the kubeadmin password to finally get access to your OpenShift cluster.

> **Important Note**
>
> kubeadmin is a temporary user with the cluster-admin privileges. It is highly recommended that you remove the kubeadmin user as soon as you set up a new identity provider, and give proper cluster-admin privileges to the cluster administrators.

Now, you can access the OpenShift Console GUI using your preferred browser. To do so, browse to https://console-openshift-console.apps.ocp.hybridmycloud.com and insert the kubeadmin credentials:

Figure 5.5 – Accessing the Console UI

Then, sit back and enjoy:

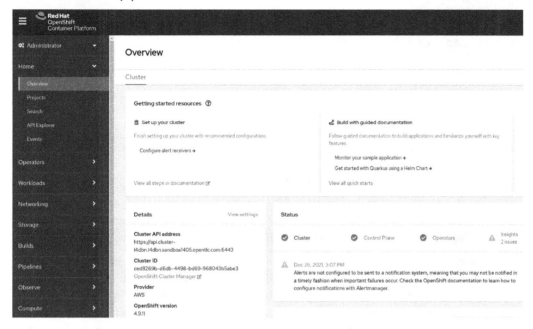

Figure 5.6 – The Console UI

Congratulations! You successfully deployed an OpenShift cluster! Celebrate your great success, but remember that your journey is just starting!

What's next?

Now, OpenShift is minimally functional, meaning that you have a control plane that can schedule pods, handle API calls and controllers, and an etcd cluster, which offers key/value storage for objects in the cluster. You also have some worker nodes fully functional that can already host some workloads.

But that's not all! Now, the activities that demand configuration begin (also known as Day 2, or post-installation activities). In the next few chapters, we will look at Ingress configurations, networks, registry configurations, views for local volumes, and persistent volumes.

We will also talk about taints and tolerations, security, and best practices – everything it takes for you to go from zero to hero and start acting on more complex implementations.

FAQs

While in the deployment phase, whatever kind of installation you do, it is common for things to not work as expected. We will look at some error cases you could face during deployment.

Why did the `openshift-install` *execution stick waiting for the API?*

The following message will be displayed:

```
Message: "INFO Waiting up to 20m0s for the Kubernetes API at
https://api.ocp.hybridmycloud.com:6443..."
```

Sometimes, the INFO message can be big trouble. At this point, you do not have much output to investigate, even if you increase the debug level of the messages. In this case, you should look at some of these options:

- Look up the API URL and check the DNS resolution; the query must result in the API IP in install-config.yaml.
- Try to ping all three master nodes individually to assure it is already up.
- UPI installation: Check on the hypervisor VM terminal if all masters are up and running and at the login prompt screen. Master nodes on the bootloader menu or whatever different situation of the login prompt screen could result in an API waiting message.
- IPI installation: Check the cloud credentials and permissions related to prerequisites. Your credentials might not have all the necessary permissions to create objects in general. Don't give admin permission to the role attributed by the cluster service user because installation searches for a specific permission name. You can check roles and permission tables at https://docs. openshift.com/container-platform/4.9/installing/installing_ vsphere/installing-vsphere-installer-provisioned.html, according to the cloud provider (for example, VMware vCenter).

Timeout installation waiting for Bootstrap to complete

The following message will be displayed:

```
Message: "INFO Waiting up to 30m0s for bootstrapping to
complete..."
```

When the openshift-install binary freezes while waiting for a Bootstrap process to complete, it means that it is waiting for some cluster operators to become available. In this case, do the following:

- Check whether you have enough worker nodes (at least two) to make the Ingress operator available.
- SSH to a worker node and check that the crictl process is still creating pods.
- SSH to a worker node and search for errors related to kube-apiserver, kubelet, podman, or crictl using the journalctl daemon.

X509 messages during cluster deployment

The following message will be displayed:

```
Message: "x509 certificate signed by unknown authority.."
```

When creating Ignition files, OpenShift automatically creates a self-signed certificate that will be verified with every API call within the cluster. However, even if you have done all the prerequisite processes properly, sometimes you may get similar `x509` messages that will result in the installation process failing and not achieving the expected result. Try checking the following options:

- **VSphere IPI installation**: Make sure you have imported the VMware CA certificate from the cluster that will extend OpenShift to the machine and will start the cluster installation.

- **Bare metal installation**: `master.ign` does not have the same CA certificate configured on the load balancer that must respond in `api-int`. Also, verify that the external load balancer has been configured to use Layer 4/TCP/Passthrough.

Certificates created by `openshift-install` residing in `master.ign` have an expiration date of 24 hours and cannot be updated. If you tried to install the day before, and it was not successful, delete the installation directory and start creating manifests and Ignition files again.

Summary

In this chapter, we have examined some options to install and configure your OpenShift Container Platform solution. From public clouds to on-premises, we have navigated through UPI, IPI, and agnostic implementation methods.

You now know about the public clouds offering fully supported implementations and documentation to start your cluster.

You are invited to look deeper into making your OpenShift cluster stronger, more reliable, and as secure as possible. We encourage you to continue to the next chapter and learn even more with us.

Further reading

If you want to look at more information related to the concepts we covered in this chapter, check the following references:

- *The installation process for disconnected installations:* `https://docs.openshift.com/container-platform/latest/installing/installing-mirroring-installation-images.html`

- *OpenShift Container Platform 4.x:* Tested integrations by Red Hat and partners: `https://access.redhat.com/articles/4128421`

- *OpenShift Container Platform IPI: x509 certificate signed by an unknown authority:* `https://access.redhat.com/solutions/5203431`

- *OpenShift Container Platform bare metal:* x509 certificate signed by an unknown authority: `https://access.redhat.com/solutions/4271572`

6

OpenShift Troubleshooting, Performance, and Best Practices

The concepts explained in *Chapter 5, OpenShift Deployment*, provided the foundation for you to initiate your first contact with an OpenShift cluster. In this chapter, we will give some tips on how to perform a health check on a cluster, dive into some **root cause analysis** (**RCA**), and also provide details on how to make a cluster run according to some best practices. Our intention with this chapter is to give you some general guidance about troubleshooting, however, it is important you always open a support ticket with Red Hat before making any changes in the platform due to troubleshooting attempts.

This chapter covers the following topics:

- Things that can crash a cluster

- Troubleshooting reference guide—how to start

- Understanding misleading error messages

> **Note**
> The source code used in this chapter is available at `https://github.com/PacktPublishing/OpenShift-Multi-Cluster-Management-Handbook/tree/main/chapter06`.

Things that can crash a cluster

Every time we start learning about some technology, it is common to be twice as careful with the installation, configuration, or adjustment to be as thorough as possible. Sometimes, to achieve these goals related to troubleshooting, performance, and best practices, the reader would turn to multiple expert readings on each of the related topics, or go through the pain of trial and error, which takes a lot of effort to succeed.

OpenShift is a great and disruptive technology, but you will navigate through a puzzle of different aspects related to storage, compute, network, and others. Obviously, in the official documentation—or even in quick internet searches—you will find commands to start from scratch, but in many situations, even with the necessary commands and parameters, it is difficult to navigate from troubleshooting to a solution.

Currently, OpenShift has an automatic recovery system, but this is usually not enough to ensure a stable environment. For this self-healing to take place successfully, many prerequisites need to be checked on the cluster first. So, before we understand what can potentially crash, let's understand how this self-adjustment mechanism works.

Operators

In the world of technology, there are many roles played, and some of them are linked to the administration of the infrastructure. There are several names for this role, the most common still being the **system administrator**, or **sysadmin**, who operates the servers and services of an **information technology** (**IT**) infrastructure. Likewise, OpenShift has *operators* that are nothing more than **applications designed to monitor platform behavior and maintain an operation**.

How do operators work? Operators are assigned to fulfill a single task of maintaining the application and all its components according to a standard. Understand that operators are not the same for all applications—that is, operators are unique, each with its own parameter definitions, configurations that are required and optional, and others.

The operator parameters' contract is described in the **Custom Resource Definition** (**CRD**). A CRD is a definition that extends a Kubernetes **application programming interface** (**API**) functionality, giving more flexibility to the cluster to store a collection of objects of a certain type. Once a CRD is defined, you can create a **Custom Resource** (**CR**) that will allow you to add a Kubernetes' custom API to the cluster.

Operators are a tool for keeping a cluster or application healthy, so why should we care about learning about OpenShift troubleshooting if it fixes itself? Indeed, operators are a powerful tool, but as we mentioned earlier, OpenShift is a big puzzle, and the pieces need to fit together perfectly for it to work properly. Although it is reigned by operators that are somehow prepared to maintain its integrity, failures can occur, and the role of the cluster administrator and their experience in solving problems will help keep all these operators healthy.

In the next sections, we'll go deeper into the main components of OpenShift and which aspects to be concerned about.

etcd

In the case of OpenShift clusters, etcd is a distributed key-value service responsible for storing the state of the cluster. Through it, all objects contained in the cluster are shown in a key-value format, so it is important to consider at least three important factors in this service, which is the heart of the control plane's operation. Note the following:

- etcd is *highly sensitive* to an infrastructure's **latency** and **bandwidth**.

- etcd needs to be distributed on all master nodes—that is, to be highly available, an OpenShift cluster infrastructure demands this service be distributed on three master nodes.

- Unlike many **high-availability (HA)** services, in which you have a main and a secondary server, with etcd, this concept is based on **quorum** members and **leadership**.

Red Hat made the etcd complexity easier by establishing the number of master nodes to be 3 as default and also by using a cluster operator that manages etcd and reports any issue in it; however, you must still understand how etcd works to be able to troubleshoot if any complex issue occurs. Go ahead to learn how the quorum and leader-based etcd algorithm works.

How do the quorum and leader-based schemes work?

An etcd cluster works on the concept of **leader** and **followers**, which is known as the **Raft Distributed Consensus** protocol. This protocol implements an algorithm based on a *leader election* to establish a distributed consensus among all members of an etcd cluster. Once members are added to an etcd cluster and a leader is elected, the process only requires sending periodic heartbeats to confirm that the leader still responds within a suitable latency time.

In case of an unanswered heartbeat time frame, the members start a new election to guarantee cluster resilience, self-healing, and continuity of service.

It is recommended that an etcd cluster has an odd number of nodes so that the following formula guarantees the tolerance of a given number of failing members. To this we give the name of **quorum**:

Quorum = (n/2)+1, where "n" represents the number of members.

A cluster must always have at least the *quorum* number of members working to be functioning properly. For the sake of clarity, let's check out some scenarios, as follows:

- **Scenario 1**: Three-member cluster, all up and running, as illustrated in the following diagram:

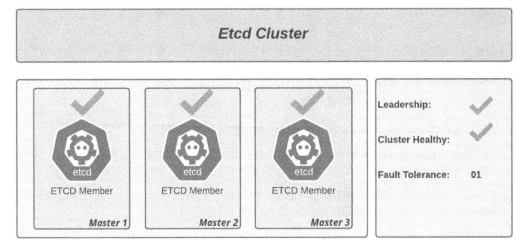

Figure 6.1 – Healthy etcd cluster (three-node member health)

Analysis: Quorum is OK as there are a majority of working members and leadership is assured, so the cluster is healthy.

- **Scenario 2**: Three-member cluster with two members working, as illustrated in the following diagram:

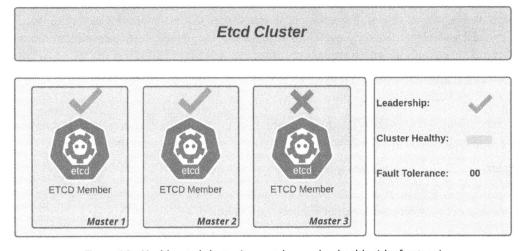

Figure 6.2 – Healthy etcd cluster (two-node member health; risk of outage)

Analysis: Quorum is OK as there are a majority of working members and leadership is assured. There is a degradation risk in case of disruption of one more node, but the cluster is healthy.

- **Scenario 3**: Three-member cluster with one member working, as illustrated in the following diagram:

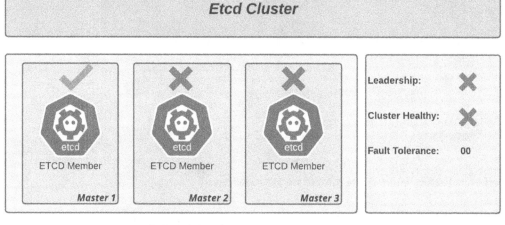

Figure 6.3 – Degraded etcd cluster (one-node member health; unhealthy cluster)

Analysis: There is no quorum as the majority of members are down, so it is no longer possible to elect a new leader, and the cluster is degraded.

Troubleshooting etcd

As mentioned earlier, OpenShift is managed by its operators, which can provide standardization, self-healing, and activities metrics, but this is not always enough to keep the cluster fully functional.

Situations such as scenarios *2* and *3* can occur due to different factors related to any infrastructure layer. It is important to carry out an in-depth analysis to restore the cluster to a functional state. Here's an approach to troubleshooting etcd.

For troubleshooting, it is important to consider what kind of disruption the cluster has been hit with. If we are still able to use the OpenShift API to access the nodes, you can use the first approach. Otherwise, if the API is unavailable, you must refer to the second scenario.

Scenario 1 – etcd member is degraded

In cases where it is still possible to execute `oc` or `kubectl` commands, use the `rsh` command to the etcd pod, and performing the following steps through `etcdctl` is the quickest approach.

An etcd degradation can occur for many different reasons, such as storage or network failures, live migration of the master nodes, or even manipulations in the **operating system (OS)**, which may cause immediate disruption of the cluster.

As mentioned before, run the following commands to open a terminal in an etcd pod and identify the failure:

```
$ oc project openshift-etcd
$ oc get pods -n openshift-etcd | grep -v etcd-quorum-guard |
grep etcd
etcd-ocp-master-0 3/3 Pending 0 14m
etcd-ocp-master-1 3/3 CrashLoopBackOff 6 17m
etcd-ocp-master-3 2/3 Running 0 9m11s
```

Notice that in the previous output, two out of three masters are getting issues, so let's `rsh` to `master-3`, perform a backup, and recreate the etcd nodes, as follows:

```
oc rsh etcd-ocp-master-3
Defaulting container name to etcdctl.
Use 'oc describe pod/etcd-ocp-master-3 -n openshift-etcd' to
see all of the containers in this pod.
sh-4.4# etcdctl member list -w table
+------+------+------+------+------+------+
| ID | STATUS | NAME | PEER ADDRS | CLIENT ADDRS | IS LEARNER
+------+------+------+------+------+------+
| 5bdacda4e0f48d91 | failed | ocp-master-1 |
https://192.168.200.235:2380 | https://192.168.200.235:2379 |
false |
| b50b656cba1b0122 | started | ocp-master-3 |
https://192.168.200.14:2380 | https://192.168.200.14:2379 |
false |
| cdc9d2f71033600a | failed | ocp-master-0 |
https://192.168.200.234:2380 | https://192.168.200.234:2379 |
false |
+------+------+------+------+------+------+
sh-4.4# etcdctl endpoint status -w table
+------+------+------+------+------+------+
| ENDPOINT | ID | VERSION | DB SIZE | IS LEADER | IS LEARNER |
RAFT TERM | RAFT INDEX | RAFT APPLIED INDEX | ERRORS |
+------+------+------+------+------+------+
| https://192.168.200.14:2379 | b50b656cba1b0122 | 3.4.9 | 136
MB | true | false | 281 | 133213554 | 133213554 | |
| https://192.168.200.234:2379 | cdc9d2f71033600a | 3.4.9 | 137
MB | false | false | 281 | 133213554 | 133213554 | |
```

```
| https://192.168.200.235:2379 | 5bdacda4e0f48d91 | 3.4.9 | 136
MB | false | false | 281 | 133213554 | 133213554 | |
+------+------+------+------+------+------+
```

Once the **identifiers (IDs)** of the failed members are determined, the next step is to remove the etcd members, leaving only the one that is running as part of the cluster. To do so, first run the backup of etcd, like so:

```
/usr/local/bin/cluster-backup.sh /home/core/assets/backup
```

> **Note**
> The etcd backup will be saved in the home directory of the core user in the same node as the one where the etcd pod is running. It is highly recommended you copy it to an off-node location as this will avoid the risk of losing access to the node and, as such, losing the etcd backup file and putting the cluster recovery at risk.

Now that you have already identified the problem with the etcd cluster, see next the suggested steps to recover your cluster.

How to solve it?

Formerly, on OpenShift version 3, the usual solution for both cases was reestablishing the etcd cluster using a backup. Now, with OpenShift 4, it is easier to provision new master nodes in your infrastructure. That said, if your issue is the first scenario (API is still available) and your installation is the **installer-provisioned infrastructure (IPI)** method, we suggest you take the following steps to recreate the problematic master node, using a new node name:

1. Get the **YAML Ain't Markup Language (YAML)** machine descriptor for any current master node by running the following command:

    ```
    oc get machine <master-node> \
        -n openshift-machine-api \
        -o yaml \
        > new-master-machine.yaml
    ```

2. The YAML file should look like this:

    ```
    new-master-machine.yaml
    apiVersion: machine.openshift.io/v1beta1
    kind: Machine
    metadata:
      finalizers:
    ```

```
   - machine.machine.openshift.io
   labels:
     machine.openshift.io/cluster-api-cluster: ocp-sgw5f
(.. omitted ..)
   name: ocp-master-4
   namespace: openshift-machine-api
   selfLink: /apis/machine.openshift.io/v1beta1/
namespaces/openshift-machine-api/machines/ocp-master-4
spec:
   metadata: {}
   providerSpec:
     value:
       apiVersion: vsphereprovider.openshift.io/v1beta1
       credentialsSecret:
         name: vsphere-cloud-credentials
       diskGiB: 120
       kind: VSphereMachineProviderSpec
(.. omitted ..)
```

3. Make the required changes in the YAML file to provision the new master, as follows:

    I. Remove the following sections or fields:

        i.   Entire status, metadata.annotations, and metadata.generation sections.

        ii. Delete metadata.resourceVersion, metadata.uid, and spec.providerId fields.

    II. Change the metadata.name field to a new name (for example, <clustername>-<clusterid>-master-3).

    III. Also update the node name in the metadata.selfLink field.

4. Delete the problematic master node using the following command:

    ```
    $ oc delete machine <problematic-master-node-name> -n
    openshift-machine-api
    ```

5. Use the following command to monitor the deletion process and certify it has been deleted:

    ```
    $ oc get machines -n openshift-machine-api -o wide
    ```

6. As soon as the problematic master has been deleted, you can now provision a new one using the YAML we prepared previously. To do so, run the following command:

```
oc apply -f new-master-machine.yaml
```

> **Note**
>
> Repeat this procedure if necessary, changing only the `metadata.name` and `metadata.selfLink` fields for each problematic master node in your cluster.

After the new master nodes have been provisioned, observe the following steps to verify whether the etcd cluster is healthy:

1. Check if all etcd pods are running, as follows. You must see three pods running:

```
$ oc get pods -n openshift-etcd | grep -v etcd-quorum-
guard | grep etcd
```

2. There are some cases in which the etcd pod is not deployed automatically with the master provisioning. If you don't see three pods running, you may run the following command to force the etcd operator to deploy the etcd pod in the new node:

```
$ oc patch etcd cluster -p='{"spec":
{"forceRedeploymentReason": "recovery-'"$( date
--rfc-3339=ns )"'"}}' --type=merge
```

3. Now, let's check from inside the etcd cluster whether it is working as expected. To do so, run the following command to open a terminal inside one of the etcd pods:

```
# Get the name of one etcd pod
$ oc get pods -n openshift-etcd | grep -v etcd-quorum-
guard | grep etcd
$ oc rsh <etcd-pod-name> -n openshift-etcd
```

4. Now, check the cluster member list, like so:

```
etcdctl member list -w table
```

5. In some cases, you will still see the problematic etcd node that we already removed as part of the cluster members. Therefore, if the previous command shows more than three members, use the following command to remove the nonfunctional etcd members:

```
$ etcdctl remove <member-id>
```

Scenario 2 – cluster API down

If the OpenShift API is down, it is important to perform any steps with much more caution to avoid an irreversible loss to the cluster. In such a scenario, you can't use the `rsh` command, get logs using `oc logs`, OR use any `oc` or `kubectl` command, as all of them use the OpenShift API, which makes troubleshooting and finding a solution much more difficult and complex.

Due to that, there must be regular etcd backups in place before the cluster malfunctions. If there is no previous backup, the first step is to perform a direct backup on the node that is in operation. To do so, proceed as follows:

- Run the following command:

    ```
    $ ssh -i ~/.ssh/id_rsa core@ocp-master-3
    ```

- Check out the etcd state by running the `crictl` command, like so:

    ```
    $ sudo crictl | grep -i etcd
    ```

- Get the etcd pod ID and run the `crictl exec` statement to identify the cluster's node state, as follows:

    ```
    $ crictl exec bd077a3f1b211 etcdctl member list -w table
    +---+---+---+---+---+---+
    | ID | STATUS | NAME | PEER ADDRS | CLIENT ADDRS | IS
    LEARNER |
    +---+---+---+---+---+---+
    | 9e715067705c0f7c | unknown | ocp-master-4 |
    https://192.168.200.15:2380 | https://192.168.200.15:2379
    | false |
    | b50b656cba1b0122 | started | ocp-master-3 |
    https://192.168.200.14:2380 | https://192.168.200.14:2379
    | false |
    | cdc9d2f71033600a | failed  | ocp-master-0
    | https://192.168.200.234:2380 |
    https://192.168.200.234:2379 | false |
    +---+---+---+---+---+---+
    ```

- Note that the etcd members are unreachable except for one that is in the started state. Run a backup by going to the node and running the backup command, as follows:

    ```
    sudo /usr/local/bin/cluster-backup.sh /home/core/assets/
    backup
    ```

- With the pod ID of etcd, obtained by `crictl`, run the following command to identify the cluster nodes and their state:

```
$ crictl exec bd077a3f1b211 etcdctl endpoint status -w
table
+--------+--------+--------+--------+--------+--------+
| ENDPOINT | ID | VERSION | DB SIZE | IS LEADER | IS
LEARNER | RAFT TERM | RAFT INDEX | RAFT APPLIED INDEX |
ERRORS |
+--------+--------+--------+--------+--------+--------+
| https://192.168.200.14:2379 | b50b656cba1b0122 | 3.4.9
| 136 MB | true | false | 491 | 133275501 | 133275501 | |
| https://192.168.200.15:2379 | 9e715067705c0f7c | 3.4.9
| 137 MB | false| false | 491 | 133275501 | 133275501 | |
+--------+--------+--------+--------+--------+--------+
```

At this stage, it is possible to draw some conclusions. We understand that there is no minimum quorum to keep the cluster up, so the API became unavailable. Continue next for a suggestion on how to proceed in this scenario.

How to solve it?

In general, to restore a cluster in this scenario you will use the command `etcdctl member remove` to remove all problematic etcd members and then use the procedure we described in the previous *How to solve it?* section to remove problematic master nodes and provision new ones. However, troubleshooting a cluster in this scenario is much more challenging, work with Red Hat support to find the best way to restore it.

Now that we have already gone through etcd troubleshooting, let's discuss another important aspect of it: performance analysis.

etcd performance analysis

Kubernetes clusters are highly sensitive to latency and throughput. Due to this, some precautions are necessary to have a stable cluster and also great performance. OpenShift is a platform designed for HA, and, as such, the expected etcd use and consumption are traffic-intensive. It is important, then, to follow some best practices to have a stable cluster. Let's look at some recommended configurations.

Storage

etcd's disk usage is intensive, so it is recommended to use **solid-state drive** (**SSD**) disks for a fast write/read response time. Regarding response times, we could say that 50 sequential **input/output operations per second** (**IOPS**) would be a minimum requirement, but from our experience, the OpenShift usage grows really fast, so we recommend you consider disks that can deliver at least 500 concurrent IOPS, to maintain the cluster's health and stability. However, note that some providers do not publish the sequential IOPS but only the shared IOPS. In such cases, consider that concurrent IOPS is equivalent to 10 times the sequential IOPS value.

Here is an example of how to measure the performance of the etcd disks using a customized version of the `fio` tool. In the OpenShift cluster, run the `debug` command to get access to a master node, like so:

```
$ oc debug node/master1.ocp.hybridcloud.com
```

As soon as the command is executed, the following message will be displayed. Execute the `chroot` command after the shell to be able to execute commands in privileged mode:

```
Starting pod/ocp-master1hybridcloud-debug ...
  To use host binaries, run `chroot /host`
  chroot /host
Pod IP: 172.19.10.4
  If you don't see a command prompt, try pressing enter.
  sh-4.4# chroot /host
  sh-4.4#
```

Create a container, as indicated in the following code snippet. After the `etcd-perf` container starts, it will automatically run performance checks:

```
sh-4.4# podman run --volume /var/lib/etcd:/var/lib/etcd:Z quay.
io/openshift-scale/etcd-perf Trying to pull quay.io/openshift-
scale/etcd-perf:latest...
Getting image source signatures
(.. omitted ..)
------- Running fio ------{
"fio version" : "fio-3.7",
"timestamp" : 1631814461,
"timestamp_ms" : 1631814461780,
"time" : "Thu Sep 16 17:47:41 2021",
"global options" : {
"rw" : "write",
```

```
 "ioengine" : "sync",
"fdatasync" : "1",
 "directory" : "/var/lib/etcd",
"size" : "22m", [1]
"bs" : "2300" }, [2]
(.. omitted ..)
"write" : {
"io_bytes" : 23066700,
"io_kbytes" : 22526,
"bw_bytes" : 1319077,
"bw" : 1288, [3]
"iops" : 573.511752, [4]
(.. omitted ..)
"read_ticks" : 3309,
"write_ticks" : 29285,
"in_queue" : 32594,
"util" : 98.318751
} ]
}
---------
99th percentile of fsync is 5406720 ns
99th percentile of the fsync is within the recommended
threshold - 10 ms, the disk can be used to host etcd [5]
```

In the preceding code snippet, we have used the following annotations:

[1] : A chunk size of 22 **megabytes** (**MB**) is usually enough to analyze performance results.

[2] : Instead of using 4k block size, etcd uses small chunks of 2.3k block size, so it guarantees performance, including with small writing fragmentations.

[3] : Bandwidth required considering traffic between the node and underlying storage. It is recommended you use at least a network interface of 1 **gigabyte** (**GB**). For medium and large clusters, the recommendation is a 10 GB interface.

[4] : The recommendation is at least 500 concurrent IOPS, as explained previously.

[5] : The report of the etcd IO check. In the example, 5.40 **milliseconds** (**ms**) demonstrates a reliable performance—for it to be so, it must be under 10 ms.

Besides using `etcd-perf` to check the disk performance, you could also perfectly use custom parameters as you need, such as block size, chunk size, and so on, using the `fio` binary tool, which is available using the standard Red Hat package manager (for example, by executing `yum/dnf install fio`).

> **Note**
> For didactic reasons, we suppressed some results, leaving only items that are pertinent to our analysis.

etcd sizing

To avoid any problems related to the **central processing unit** (**CPU**), you must understand whether your cluster is well-sized. You must consider some factors to check the cluster sizing, such as the number of customers using the platform, the expected number of requests per second, and the amount of storage available for etcd.

First, let's give you some parameters to consider for your cluster size:

Cluster Size	Nodes	Clients	Requests/s	Store
Small	Up to 50	Up to 100	Up to 200	Up to 100 MB
Medium	Up to 250	Up to 500	Up to 1000	Up to 500 MB
Large	Up to 1000	Up to 1500	Up to 10000	Up to 1024 MB
Extra Large	Up to 3000	Over 1500	Over 10000	Over 1024 MB

The following table demonstrates some use cases using public clouds and on-premises infrastructures according to the amount of CPU, memory, disk IOPS, and bandwidth linked to the cluster size:

Cluster Size	Provider	Virtual CPU (vCPU)	Memory	IOPS	Disk Bandwidth (MB/s)
Small	Amazon Web Services (AWS)	2	8 GB	3600	56.25
Small	Google Cloud Platform (GCP)	2	7.5 GB	1500	25
Small	On-premises	4	16 GB	300	N/A
Medium	AWS	4	16 GB	6000	93.75
Medium	GCP	4	15 GB	4500	75
Medium	On-premises	4	16 GB	300	N/A
Large	AWS	8	32 GB	8000	125
Large	GCP	8	30 GB	7500	125
Large	On-premises	8	16 GB	300	N/A
Extra Large	AWS	16	64 GB	16000	250
Extra Large	GCP	16	64 GB	15000	250
Extra Large	On-premises	16	16 GB	300	N/A

In a nutshell, when you size a cluster, you should consider these thresholds because this is already benchmarked by the etcd community, and their performance will likely be acceptable if these recommendations are followed. Further information regarding sizing the etcd cluster can be found at the link we have provided in the *Further reading* session of this chapter.

In this section, you have seen some ways to check etcd performance and troubleshoot, and you also got some important information regarding sizing best practices. We hope you enjoyed the approach and take a close look at the next section about authentication, which will be another interesting theme.

Authentication

Another important aspect of an OpenShift cluster is user authentication and authorization flow. OpenShift's flexibility and easy-to-use authentication plugins are a smart way of setting up users and groups. Instead of simply having a vault of usernames and passwords, OpenShift's authentication service can authenticate a user in a variety of ways—we call it an **identity provider** (**IdP**). In this way, OpenShift is responsible for trusting the IdP and allowing or denying authentication according to the provider. In the following diagram, you can see how the process of authenticating a user works:

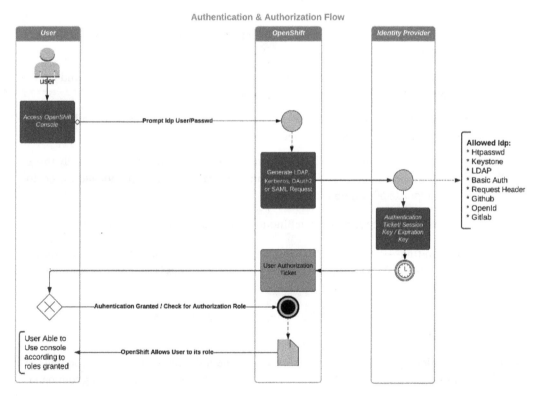

Figure 6.4 – Authentication and authorization flow

The IdP is responsible for notifying OpenShift if the username and password are valid and returning to OpenShift the `success` or `failure` authentication status. This process is known as **authentication** (**AuthN** in some literature).

We mentioned that the authentication process uses IdPs to validate the user against an authentication provider, but, more importantly, you need to understand how **authorization** (aka as **AuthZ**) occurs. Initially, a user on OpenShift doesn't have any permissions in any project; however, they can log on. They will not have rights to perform any tasks except creating their own new projects if the cluster has the self-provisioner role enabled (self-provisioner is a role that allows any logged user to create their own projects). To add permissions to a user, it is necessary to inform the appropriate role. This can be accomplished through the inclusion of a user group or by directly assigning it to the user. This process of adding roles to users or groups is named **RoleBindings**.

To better understand which role best fits a user or group, consider looking at the default roles that already exist in an OpenShift cluster, as set out here:

- `admin`—This is a project admin. It allows changes with all project-scoped resources, including the ability to create RoleBindings using ClusterRoles. It does not allow the modification of quotas, limits, or cluster resources.

- `edit`—This is a project editor. It allows usage and manipulation of all project-scoped resources, but cannot change authorization objects.

- `view`—This is a project viewer. It allows the inspection of project-scoped resources and works like a read-only RoleBinding. Secrets inspections are not allowed.

- `cluster-admin`—This is the equivalent of a root user on a *nix-like OS. It allows total control of any resource in the entire cluster.

- `cluster-reader`—This is useful for allowing special users permissions, especially those that work with cluster monitoring. This RoleBinding is read-only and does not allow users to escalate or manipulate objects on the cluster.

You must also understand the scope of the RoleBinding, which can be one of the following:

- **Local RoleBinding**—Permissions that are given in a specific project; for example, adding the `edit` role to user *X* in project *Z*. Creating local RoleBindings is as simple as running the following command:

  ```
  $ oc adm policy add-role-to-user <role> <user> -n
  <project>
  ```

- **Cluster RoleBinding**—Permissions that are given for the entire cluster; for example, adding the `cluster-admin` role to user *X*. To create a cluster RoleBinding, run the following command:

  ```
  $ oc adm policy add-cluster-role-to-user <role> <user>
  ```

Similar commands can be applied to groups, just replacing user with group (for example, add-role-to-group and add-cluster-role-to-group). Similarly, to remove a role from a user or group, use remove-role-from-user/group, as in the following example:

```
$ oc adm policy remove-role-from-user <role> <user> -n
<project>
$ oc adm policy remove-role-from-group <role> <group> -n
<project>
$ oc adm policy remove-cluster-role-from-user <role> <user>
$ oc adm policy remove-cluster-role-from-group <role> <group>
```

These are some of the most popular default roles that are used with OpenShift, but you can create custom ones if needed. To create a custom role, you need to first understand what are **verbs** and **resources**, so here are definitions of these:

- A **verb** is an action the user runs against the OpenShift API—for instance, get, list, create, update, and so on.

- A **resource** is an entity in which the verb will be performed—for example, pod, deployment, service, secret, and so on.

That said, to define a custom role, you need to know which verbs for the user or group will be allowed to run over which objects. As soon as you have defined verbs and resources, a role can be created using the following command:

```
$ oc create role <role-name> --verb=<verbs-list>
--resource=<resources-list>
```

Have a look at the following example:

```
$ oc create role sample --verb=get,list,watch
--resource=pods,pods/status
```

There are more things about authentication and authorization with OpenShift that it's not our intention to bring to light here. We tried to highlight some of the important aspects you need to know about it, and we left a set of links in the *Further reading* section of this chapter if you want to go even deeper into this subject.

With that, we have demystified the authentication process a bit, and you can now perform the process of **AuthN** and **AuthZ**. The previous diagram showed a quick point of view about the steps of an authentication process. It is important to give proper permissions to each user or group, and—more importantly—to plan the roles you will need to have in place to give your users and groups the proper permissions to perform their job. In the following section, we will cover another important aspect that an OpenShift operator needs to know about: troubleshooting.

Troubleshooting reference guide – how to start

In this section, you will see some approaches to troubleshooting your OpenShift cluster if you face any issues. Due to the power of the oc **command-line interface** (**CLI**), you will have different ways to succeed in almost any troubleshooting scenario of your OpenShift cluster. Along with your training, you will gain the experience you need to take a step further in using and troubleshooting your OpenShift/Kubernetes cluster.

Describing objects

As we have mentioned, the oc CLI is a powerful tool to help OpenShift users to do a lot of operations and also do some troubleshooting. One of the first steps of troubleshooting is to get some details and descriptions of the objects. Suppose, for instance, you have an issue related to a pod that is not coming up, for some reason. Let's start our troubleshooting by checking the pod details, as follows:

```
$ oc describe pod sso-10-qm2hc
```

Check the output in the Events section of the object to see what is preventing the pod from spinning up, as illustrated in the following code snippet:

```
Events:
  Type    Rea-
son         Age                     From              Message
  ----    ------             ----                 ---
-           -------
  Normal  Scheduled          90s                  default-schedul-
er  Successfully assigned rhsso/sso-10-qm2hc to ocp-hml4.hy-
bridcloud.com
  Normal  AddedInterface  89s                 mul-
tus            Add eth0 [10.242.22.12/23] from openshift-sdn
  Normal  Pulling            39s (x3 over 89s)   kube-
let         Pulling image "image-registry.openshift-im-
age-registry.svc:5000/rhsso/sso74-custom"
  Warning Failed             33s (x3 over 83s)   kube-
let           Failed to pull image "image-registry.open-
shift-image-registry.svc:5000/rhsso/sso74-custom": rpc error:
code = Unknown desc = pinging container registry image-reg-
istry.openshift-image-registry.svc:5000: Get "https://im-
age-registry.openshift-image-registry.svc:5000/v2/": dial tcp
10.244.109.169:5000: connect: no route to host
```

```
   Warning   Failed           33s (x3 over 83s)   kube-
let             Error: ErrImagePull
   Normal    BackOff          8s (x4 over 83s)    kube-
let             Back-off pulling image "image-registry.open-
shift-image-registry.svc:5000/rhsso/sso74-custom"
   Warning   Failed           8s (x4 over 83s)    kube-
let             Error: ImagePullBackOff
```

In this case, you were able to see quickly in the `oc describe` command that the error is related to the connection between the node and the image registry (`no route to host`). You can act accordingly to fix the connectivity issue and get the pod up and running. You can also use the `Events` log to see other meaningful information, as you can see in the following section.

Events

Another parameter on the `oc` CLI that helps with problem investigation is the `oc get events` command. This is very useful for showing a log of tasks recently executed, along with presenting *success* or *error* messages. Events can be executed cluster-wide or project scoped. Check out the following sample of event logs:

```
$ oc get events -n openshift-image-registry
LAST SEEN    TYPE       REASON          OB-
JECT                                            MESSAGE
35m          Normal     Scheduled       pod/cluster-im-
age-registry-operator-7456697c64-88hxc    Successfully as-
signed openshift-image-registry/cluster-image-registry-opera-
tor-7456697c64-88hxc to ocp-master2.hybridcloud.com
35m          Normal     AddedInterface       pod/clus-
ter-image-registry-operator-7456697c64-88hxc    Add eth0
[10.242.0.37/23] from openshift-sdn
35m          Normal     Pulled               pod/cluster-im-
age-registry-operator-7456697c64-88hxc    Container image
"quay.io/openshift-release-dev/ocp-v4.0-art-dev@sha256:6a-
78c524aab5bc95c671811b2c76d59a6c2d394c8f9ba3f2a92bc05a780c783a"
already present on machine
(...omitted...)
```

If the pod is up and running but you still have some issues in an application, you can also use OpenShift to check the application logs, as you will see next.

Pod logs

Regularly, pod logs bring important information related to the scheduler, pod affinity/anti-affinity, container images, and persistent volumes. There are several ways to check the pod logs, as outlined here:

- The common way—inside the namespace—is shown here:

```
$ oc project mynamespace
$ oc logs mypod
```

- Here's how to check them from any namespace:

```
$ oc -n mynamespace logs mypod
```

- Here's how to check the logs of a specific container inside a pod:

```
$ oc -n mynamespace logs mypod -c kube_proxy
```

- You can also check the logs using the OpenShift console **user interface (UI)**. To do so, access the Logs tab of the desired namespace and pod, as illustrated in the following screenshot:

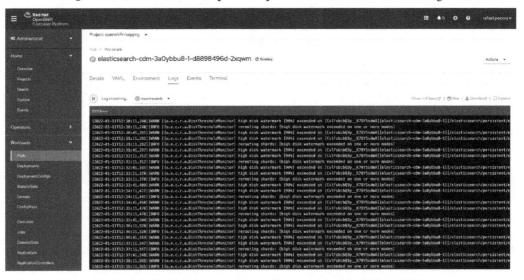

Figure 6.5 – Pod logs example (OpenShift console graphical UI (GUI))

You may also have issues during the application deployment. See next what you can check to find evidence of deployment problems.

Deployment logs

In some cases, a pod does not start and remains in a constant crashing state, which makes it difficult to get any logs. In such cases, you can check the `deployment` or `deploymentconfig` logs, which can help you to identify deployment misconfigurations.

Similar to pod logs, `deployment` logs are accessible by running the `oc logs` command. For deployment logs, run the following command:

```
$ oc -n mynamespace logs deployment/mydeploypods
```

For `deploymentconfigs` logs, use this one:

```
$ oc -n mynamespace logs dc/mydeploypods
```

Usually, you will not find the exact issue or the root cause and the solution to be applied, but it will give you a good indication of why it is failing—for example, dependent components that are missing in the solution, such as images not available; security context constraint with incorrect permissions; configmaps, secrets, and serviceaccounts missing; and so on.

Another useful way you may use to troubleshoot an issue is to use the `debug` command with temporary root privileges. Learn more on this in the following section.

Debugging pods

Another interesting tool to troubleshoot a pod that is constantly crashing can be used by executing the `oc debug deployment` or `oc debug deploymentconfig` command. Through this, you can instruct OpenShift to not fail and restart the pod as it crashes. The pod will still be alive, so you can check the logs, access it, and troubleshoot it from inside the container. To use the tool, run the following command:

```
$ oc debug deployment/<deployment-name>
```

> **Note**
>
> The `oc debug` command allows some interesting options, such as `--as-user` to run the pod with a defined user. To see a comprehensive list of allowed parameters and examples, run the `oc debug -h` command.

Operator logs

As we already covered in this book, OpenShift uses several operators to deploy and monitor critical features for the platform. That said, operators bring helpful logs to identify some configuration issues and instability with the platform. Those logs are stored in namespaces starting with the name `openshift-*`, which is a standard for most tools included with OpenShift.

The reason to maintain several operator pods as a part of project pods is related to some affinity/anti-affinity rules and taints/toleration strategies that a user can apply to the cluster. Operator pods are the watchdog to maintain namespaces' health, watching the CRD, its standards, liveness, and readiness, and preventing OpenShift's critical namespaces from suffering undesirable changes.

One of the main benefits of operators for the cluster's stability is the ability to maintain the desired state of the operator namespace's objects. In other words, if any unwanted changes are made by mistake directly into the namespace's objects, they will be reverted by the operator itself. Every change in the operator's objects needs to be done through the operator itself, by editing ConfigMaps or CR objects, according to the operator's specification. That also means that any changes are checked and confirmed by operators before they are effectively applied.

To check the cluster operators' functions, you must do the following:

1. List the cluster operators, like so:

    ```
    $ oc get co
    ```

2. Describe the details of the cluster operator, as follows:

    ```
    $ oc describe co <clusteroperatorName>
    ```

    Here's an example:

    ```
    $ oc describe co storage
    ```

3. Check the status output for error messages, if they exist, like so:

    ```
    (...omitted...)
    Status:
      Conditions:
        Last Transition Time:  2021-08-26T14:51:59Z
        Message:               All is well
        Reason:                AsExpected
        Status:                False
        Type:                  Degraded
        Last Transition Time:  2021-08-26T14:51:59Z
        Message:               All is well
        Reason:                AsExpected
        Status:                False
        Type:                  Progressing
        Last Transition Time:  2021-08-26T14:51:59Z
        Message:               DefaultStorageClassControl-
    ```

```
lerAvailable: No default StorageClass for this platform
    Reason:                 AsExpected
    Status:                 True
    Type:                   Available
    Last Transition Time:   2021-08-26T14:52:00Z
    Message:                All is well
    Reason:                 AsExpected
    Status:                 True
    Type:                   Upgradeable
(...ommitted...)
```

In the previous example, only a warning message about setting up a default storage class for a cluster is shown. No critical issues were found at storage. In case of any issue, look at the events and pod logs in the operator's namespace.

Other oc CLI commands and options

The oc CLI has some other powerful commands for troubleshooting. Some useful and powerful commands for troubleshooting are listed here:

- Here's an example command for high-verbosity logs:

  ```
  $ oc -n <namespaceName> logs <podName> -v 8
  ```

- And here's one for cluster events:

  ```
  $ oc get events
  ```

- Here, you can see an example command for namespaced events:

  ```
  $ oc -n <namespaceName> get events
  ```

- Here's how to execute a single command inside the pod (double dash required):

  ```
  $ oc exec mypod -- date
  ```

- Here's how to execute a single command inside a pod in a specific container (double dash required):

  ```
  $ oc exec mypod -c httpd-container -- date
  ```

- Create iterative commands spawning a pseudo-terminal, like so:

```
$ oc exec mypod -i -t -- ls -t /usr
```

- Similar to the exec command, rsh—shown here—opens a terminal inside the pod:

```
$ oc -n <namespaceName> rsh  <podName>
```

All of the previous commands help you to identify problems in a cluster or even an application. Furthermore, you can investigate the node directly using the oc debug command, like so:

```
$ oc debug node/<nodeName>
```

The oc debug command gives you non-root privilege access, and you cannot run many OS commands without escalation. To do so, we recommend you run the chroot command, like so. After that, you can regularly use OS shell commands:

```
$ chroot /host /bin/bash
```

As you can see, OpenShift has a lot of useful debugging commands to help you identify cluster-wide or scoped issues. It is not recommended, but it is possible to also directly ssh on nodes. This kind of approach requires good knowledge of the *Red Hat CoreOS* OS, podman, and crio to avoid node disruption.

In any situation, we also recommend you open a support ticket with Red Hat, which will assist and give you the right guidance to solve your problem. The Red Hat support team often asks for the result of the must-gather command, which generates a temporary pod and concatenates meaningful logs and configurations that are useful for the Red Hat engineering team to analyze and correlate events and find the issue or root cause.

The most common way to run must-gather is shown here:

```
$ oc adm must-gather --dest-dir=/local/directory
```

This will create a tar file under the chosen directory with all logs collected, which can be very useful to identify problems. We suggest you always run this command and upload it when you are opening the support ticket to speed up the process of analyzing the issue.

In this section, you saw different debug approaches that will certainly help you in everyday life. In the next section, you will see the most common error messages that occur at pod startup, and in this way, you will be able to draw your own line of reasoning that will help you in the problem solution.

Understanding misleading error messages

Even if you have learned the different ways to identify a problem, it is not unusual that the error shown does not provide enough information to help you to detect the issue and fix it. Having that in mind, we decided to highlight some very common error messages in this section and also some suggestions to solve the problem.

ImagePullBackOff

This is a common error related to a missing container image. Check out the following lines of code to become familiarized with this kind of issue when you face it:

```
NAMESPACE    NAME READY STATUS RESTARTS AGE
namespace1   backend-tfmqm 0/1 ImagePullBackOff 0 17h
```

Here's a message that may come up when investigating the pod log:

```
$ oc -n namespace1 logs backend-tfmqm
Error from server (BadRequest): container " backend" in pod "
backend-tfmqm" is waiting to start: trying and failing to pull
image
```

Looking at the error message, it is typically linked to the absence of the image in the registry. This can occur due to some problems, such as the image and its tags not being available in the registry, or incorrect pointing in the **deployment/deployment config**. Another correlation would be the node where the pod is running is not able to reach the image registry.

CrashLoopBackOff

This is an error that requires some knowledge before acting on solving it effectively. It occurs because the application crashes constantly, so the root cause issue can be due to several different reasons. You can see an example of this here:

```
NAMESPACE NAME READY STATUS RESTARTS AGE
3scale backend-redis-1-9qs2q 0/1 CrashLoopBackOff 211 17h
```

Here's a message you may see when investigating the pod log:

```
$ oc logs backend-redis-1-9qs2q
1:M 11 Jan 13:02:19.042 # Bad file format reading the append
only file: make a backup of your AOF file, then use ./redis-
check-aof --fix <filename>
```

The log usually gives you some hints about the issue's root cause, but it can also be a trap to conduct you to a wrong conclusion. It is important to take into account when a pod has a persistent volume or it has a precedence order that depends on another application to start first and prepare a persistent volume, as well as many other different scenarios that can lead to this error.

Init:0/1

When you get an `Init:0/1` error message, it typically means the pod is waiting for another pod or a condition that hasn't been satisfied yet. The following lines of code demonstrate what kinds of conditions can result from this message and how to solve this:

```
NAMESPACE NAME READY STATUS RESTARTS AGE
3scale backend-cron-1-zmnpj 0/1 Init:0/1 0 17h
```

Here's a message you may see when investigating the pod log:

```
$ oc logs backend-cron-1-zmnpj
Error from server (BadRequest): container "backend-cron" in pod
"backend-cron-1-zmnpj" is waiting to start: PodInitializing
```

This, perhaps, can be a confusing status when you are troubleshooting error messages. Certainly, it can be anything wrong in the namespace, so the error message only shows `PodInitializing`. You could interpret it as a condition when a pod is waiting to start; meanwhile, this message means that a condition isn't satisfied.

To help you, we have listed here some items that must be checked that may be preventing the pod from starting:

- Check whether the service account used in the pod exists in the namespace, as some containers need a specific service account name and policy applied to start.

- Check the **security context constraints** (**SCCs**) and make sure that these are properly set according to the required permissions for the pod.

- Check other containers and pods in the namespace: depending on the build strategy, it is possible to define pod dependencies.

If you are not familiar with SCC yet, don't freak out. We will be covering it in depth in *Chapter 8, OpenShift Security*.

Summary

In this chapter, we focused on important components of OpenShift, such as operators and their role in maintaining the resilience of a cluster, and we also talked about situations that can cause damage to the cluster.

We have dived into the heart of OpenShift, which is its distributed database (known as etcd), understanding its importance in the cluster and how to prepare it to receive a high volume of traffic, as well as verifying its sizing and performance, and understanding how to perform troubleshooting in some cases.

We have also discussed a bit about the AuthN and AuthZ process, so you now know the power and flexibility of the Openshift IDPs. We finally have seen some important troubleshooting tips and tools that will certainly help you in your daily job, operating OpenShift clusters and applications.

In the next chapter, we will present some other important information about the network on OpenShift. We will discuss and give examples to understand the main differences between a pod network and a service network, as well as understand the difference between North-South and East-West traffic. Keep up with us in this interesting reading and learn more in *Chapter 7, OpenShift Network*.

Further reading

If you want to look at more information on what we covered in this chapter, check out the following references:

- *Red Hat Knowledgebase—etcd recommendation*: `https://access.redhat.com/solutions/4770281`
- *etcd quorum model*: `https://etcd.io/docs/v3.5/faq/`
- *Understanding etcd quorum*: `http://thesecretlivesofdata.com/raft/`
- *etcd hardware sizing recommendations*: `https://etcd.io/docs/v3.5/op-guide/hardware/`
- *etcd tuning options*: `https://etcd.io/docs/v3.5/tuning/`
- *etcd benchmarking thresholds*: `https://etcd.io/docs/v3.5/benchmarks/`
- *etcd benchmark CLI tool*: `https://etcd.io/docs/v3.5/op-guide/performance/#benchmarks`
- *Kubernetes authentication flow*: `https://kubernetes.io/docs/reference/access-authn-authz/authentication/`
- *Openshift IDPs*: `https://docs.openshift.com/container-platform/4.7/authentication/understanding-identity-provider.html`

- *Learn more about SCCs*: `https://docs.openshift.com/container-platform/4.8/authentication/managing-security-context-constraints.html`

- *More information about troubleshooting*: `https://docs.openshift.com/container-platform/4.7/support/troubleshooting/investigating-pod-issues.html`

- *Recommended etcd practices*: `https://docs.openshift.com/container-platform/4.10/scalability_and_performance/recommended-host-practices.html#recommended-etcd-practices_recommended-host-practices`

- *How to calculate IOPS in a storage array (blog article)*: `https://www.techrepublic.com/article/calculate-iops-in-a-storage-array/`

7
OpenShift Network

As we know, networking can be the cause of big trouble if it is not well designed. From a traditional perspective, the network is the dorsal spine of every infrastructure. Networking equipment such as routers, modems, switches, firewalls, **Web Application Firewalls (WAFs)**, **Intrusion Detection Systems/ Intrusion Prevention Systems (IDSs/IPSs)**, proxies, and **Virtual Private Networks (VPNs)** needs to be totally integrated, deployed, and maintained using best practices to ensure high performance and reliable network infrastructure. In this chapter, we will discuss important concepts related to networking on OpenShift that you need to take into consideration to make the best decisions for your case.

This chapter covers the following topics:

- OpenShift networking
- Network policies
- What is an Ingress controller?
- Types of routes

OpenShift networking

Throughout this book, we continue to reaffirm the importance of choosing the right architecture as it directly impacts the way the cluster will work. We expect that, at this time, all the required network decisions have been made and implemented already – there are a lot of network changes that are not possible after cluster deployment.

Although we already discussed networks in *Chapter 2, Architecture Overview and Definitions*, and deployed our cluster, we believe that it is important to expand on this topic a bit more and include more details about the differences when considering network usage.

Red Hat OpenShift uses a default **Software-Defined Network** (**SDN**) based on Open vSwitch (`https://github.com/openvswitch/ovs`) that creates a multilayer network solution. This additional layer works as a virtual switch on top of the network layer, and it is responsible for creating, maintaining, and isolating traffic on the virtual LAN.

Because of its multiple-layer network capacity, Open vSwitch provides a way to control traffic coming in and out of the cluster. Refer to the following diagram to better understand network traffic between network layers:

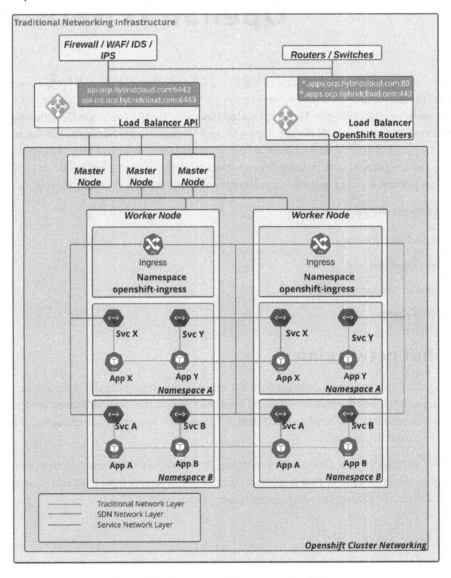

Figure 7.1 – Overview of the networking layers

During the OpenShift cluster installation, some namespaces related to network functions are created; basically, the most important network project is openshift-sdn, which contains some pods for each node that will be responsible for the traffic between the nodes. It is relevant to also state that the traffic is running inside a virtual LAN operated by Open vSwitch. There are other network projects involved as well, such as openshift-host-network and openshift-ingress.

How does traffic work on Open vSwitch?

To answer this question, we need to define where the traffic begins. Let's start with the internal traffic, which means the communication between the application's pods that are inside the OpenShift cluster.

To facilitate your understanding, consider two applications running on OpenShift; the first one is named app-frontend and the second app-backend. As the name suggests, app-frontend makes API calls to app-backend to process user requests.

Therefore, when a pod from the app-frontend application makes a request to the app-backend application, this request will be sent to the internal service, in this case the app-backend service. The app-backend service is responsible for delivering that package to one of the app-backend pods. In the same way, the application handles its request and sends the result package back to the service network, which, at this point, already has a connection established with app-frontend.

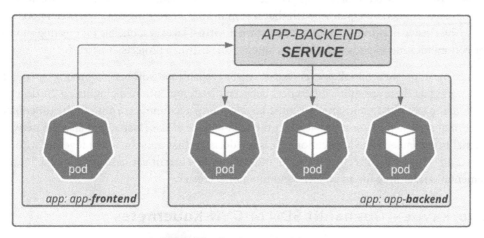

Figure 7.2 – Service network layer

With that, we have briefly explained the traffic between applications inside the cluster. Now, let's see how external-to-internal traffic is handled. When a request comes from outside the cluster, it goes initially to the external load balancer. As the load balancer receives a connection, it routes the request to one of the *OpenShift Ingress* pods, which sends it to the service of the destination application, which, in turn, routes it to the proper application's pod.

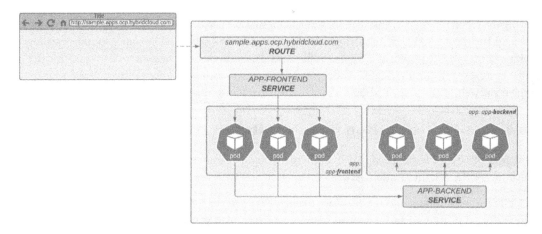

Figure 7.3 – Route SDN networking

Now that you understand how traffic works in an OpenShift cluster, it is important to reinforce that OpenShift basically works with three network layers: the node network, the service network, and the cluster network (aka the pods network).

The **node network** is the physical network used to create and maintain machines. The **service network** is a virtual layer created by Open vSwitch that is responsible for routing traffic between pods and services. The **cluster network** is another Open vSwitch virtual layer responsible for creating subnets for the communication of pods – it allows isolating traffic between projects as needed.

In the next sections, we will look deeper into the main available networking plugins for OpenShift. Keep in mind that there are subtle differences between the aforementioned plugins, so the decision between using one plugin and another must be taken into account according to the differences in functionality, which can somewhat affect the architecture of the cluster, and also the network functionality available to the applications. This is a decision that must be made together with the network and software architecture team, to understand the current use cases and planned future implementations, aiming for an efficient and functional cluster.

Network type – OpenShift SDN or OVN-Kubernetes

OpenShift is a complete PaaS solution based on Kubernetes that provides several options other than its default components. For instance, OpenShift, by default, uses the Open vSwitch network plugin (OpenShift SDN), but you can use **OVN-Kubernetes** as an alternative.

A network plugin is a feature that creates an overlay network using the Kubernetes **Container Network Interface** (**CNI**) that isolates the traffic between the virtual machines network and the OpenShift nodes.

These two supported options offer a good and reliably performing network, but you can use other kinds of CNI depending on the scenario where OpenShift has been provisioned. Check the link for *OpenShift Tested Integrations* in the *Further reading* section of this chapter to see the options that are tested and supported by Red Hat.

Network policies

As we already mentioned, OpenShift uses an SDN, and preferably, the network traffic control should be done using the features the cluster provides itself. In our experience, having implemented OpenShift in many organizations, we have often heard doubts regarding how to control network traffic within the cluster, as most customers are used to doing it by using regular firewall devices. In this section, we will walk you through how to control network traffic to be able to allow or deny network traffic as needed. Before giving you some options to do that, we first need to differentiate the different traffic directions that we have in a cluster.

North-south traffic

OpenShift has been designed to cover the most common scenarios, even regarding networking. When an incoming connection comes from outside the cluster to an application, it is possible to control network traffic into the cluster using an external firewall and/or the **OpenShift Ingress** solution.

East-west traffic

Initially, it may sound a little weird to say that there is also network traffic in east-west directions but east-west network traffic is nothing more than traffic between applications in different namespaces inside the same OpenShift cluster.

The following diagram explains how these different types of traffic occur in a cluster:

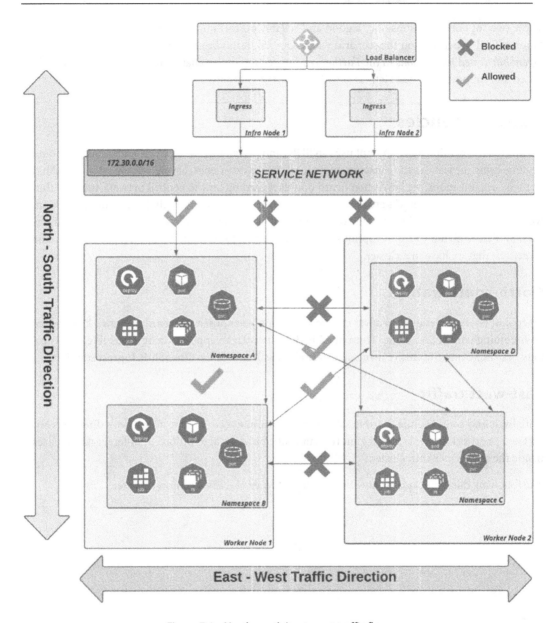

Figure 7.4 – North-south/east-west traffic flow

You have seen the possible directions in which the traffic on the network can be controlled. In the next section, you will see how to control the network traffic in the cluster.

Controlling network traffic

There are different options for controlling traffic on OpenShift:

- For north-south traffic, you can either use an external firewall and load balancer to control the traffic before getting into the OpenShift cluster or use annotations in the OpenShift **route** object to control aspects such as the rate limit, timeout, and load balancing algorithm.

- Use a proper **network policy** to allow or deny a traffic flow, as needed.

- Use the **ovs-multitenant** network isolation mode. This mode was commonly used on OpenShift version 3 but is not encouraged on version 4, as the Network Policy plugin has become the standard.

- If you intend to use microservices with OpenShift, you may also choose to use a **service mesh** to control the east-west traffic, which uses the **istio-proxy** sidecar to give the lowest granularity of isolation mode. Service meshes are not the focus of this book, but if you want more information on them, check out the *Further reading* section of this chapter.

> **Note**
>
> If you used to use **ovs-multitenant** on OpenShift 3.x and want to have similar functionality on version 4.x, we recommend you customize the project template, adding network policies to block traffic between different projects by default. The process to do that is simple and described at this link: `https://docs.openshift.com/container-platform/latest/networking/network_policy/default-network-policy.html`.

In this chapter, we will focus on Network Policy, as this is the standard network plugin on OpenShift 4. See next how to create a network policy to control the network traffic.

Creating a network policy

As we already mentioned, with network policies, you can define rules to allow or block ingress network traffic in a cluster. With a network policy, you can, for instance, allow traffic between pods inside the same namespace but deny it from other namespaces. You may also allow traffic only on a specific port, and so on. Therefore, for a better understanding of network policies and the directions in which traffic is and isn't allowed to flow, we will provide several diagrams and scenarios to clarify the importance of namespace isolation.

For learning purposes, we will use three namespaces, named `bluepets`, `greenpets`, and `otherpets`. In the following diagram, we are illustrating the default **network policy**, which allows traffic between namespaces and traffic from a cluster ingress by default:

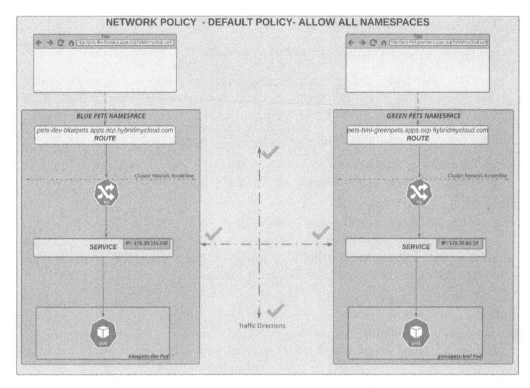

Figure 7.5 – Default network policy – allow all

So, let's go ahead and demonstrate connections allowed to these two namespaces: bluepets and greenpets. To facilitate your understanding, we are running tests in an external network with no direct route to the **service network**, which is the only routable network from our cluster. So, to simulate all scenarios, we access the pod using rsh on the greenpets namespace and try to reach the service IP of the bluepets namespace in our lab scenario discussed previously.

Before going into that, we must get the service IPs from both the namespaces to use later in the pod terminal and check the results accordingly.

```
% oc get svc -n bluepets
NAME       TYPE        CLUSTER-IP      EXTERNAL-IP   PORT(S)                      AGE
pets-dev   ClusterIP   172.30.116.236  <none>        8080/TCP,8443/TCP,8778/TCP   16m
                                                     pecs-cluster ~ (greenpets) - 0:25:21]
% oc get svc -n greenpets
NAME       TYPE        CLUSTER-IP      EXTERNAL-IP   PORT(S)                      AGE
pets-hml   ClusterIP   172.30.110.60   <none>        8080/TCP,8443/TCP,8778/TCP   12m
                                                     pecs-cluster ~ (greenpets) - 0:26:06]
% oc get networkpolicy -n bluepets
No resources found in bluepets namespace.
                                                     pecs-cluster ~ (greenpets) - 0:26:34]
% oc get networkpolicy -n greenpets
No resources found in greenpets namespace.
```

Figure 7.6 – Service IPs – bluepets and greenpets namespaces

Take a look at the following screenshot. We `rsh` a pod under the `greenpets` namespace and run `curl` on the following endpoints:

- The service IP in `greenpets` (the same namespace): To check connectivity between a pod and service in the same namespace (highlighted with a green square in the following screenshot).

- The service IP in `bluepets` (a different namespace): We similarly call the service IP of the `bluepets` namespace and it also works fine (highlighted with a blue square in the following screenshot).

```
sh-4.4$ curl -Iv http://172.30.110.60:8080
* Rebuilt URL to: http://172.30.110.60:8080/
*   Trying 172.30.110.60...
* TCP_NODELAY set
* Connected to 172.30.110.60 (172.30.110.60) port 8080 (#0)
> HEAD / HTTP/1.1
> Host: 172.30.110.60:8080
> User-Agent: curl/7.61.1
> Accept: */*
>
< HTTP/1.1 200
HTTP/1.1 200
< Content-Type: text/html;charset=UTF-8
Content-Type: text/html;charset=UTF-8
< Content-Language: en
Content-Language: en
< Content-Length: 2872
Content-Length: 2872
< Date: Thu, 17 Mar 2022 16:12:34 GMT
Date: Thu, 17 Mar 2022 16:12:34 GMT

<
* Connection #0 to host 172.30.110.60 left intact
sh-4.4$
sh-4.4$
sh-4.4$
sh-4.4$
sh-4.4$
sh-4.4$ curl -Iv http://172.30.116.236:8080
* Rebuilt URL to: http://172.30.116.236:8080/
*   Trying 172.30.116.236...
* TCP_NODELAY set
* Connected to 172.30.116.236 (172.30.116.236) port 8080 (#0)
> HEAD / HTTP/1.1
> Host: 172.30.116.236:8080
> User-Agent: curl/7.61.1
> Accept: */*
>
< HTTP/1.1 200
HTTP/1.1 200
< Content-Type: text/html;charset=UTF-8
Content-Type: text/html;charset=UTF-8
< Content-Language: en
Content-Language: en
< Content-Length: 2872
Content-Length: 2872
< Date: Thu, 17 Mar 2022 16:12:50 GMT
Date: Thu, 17 Mar 2022 16:12:50 GMT

<
* Connection #0 to host 172.30.116.236 left intact
sh-4.4$
```

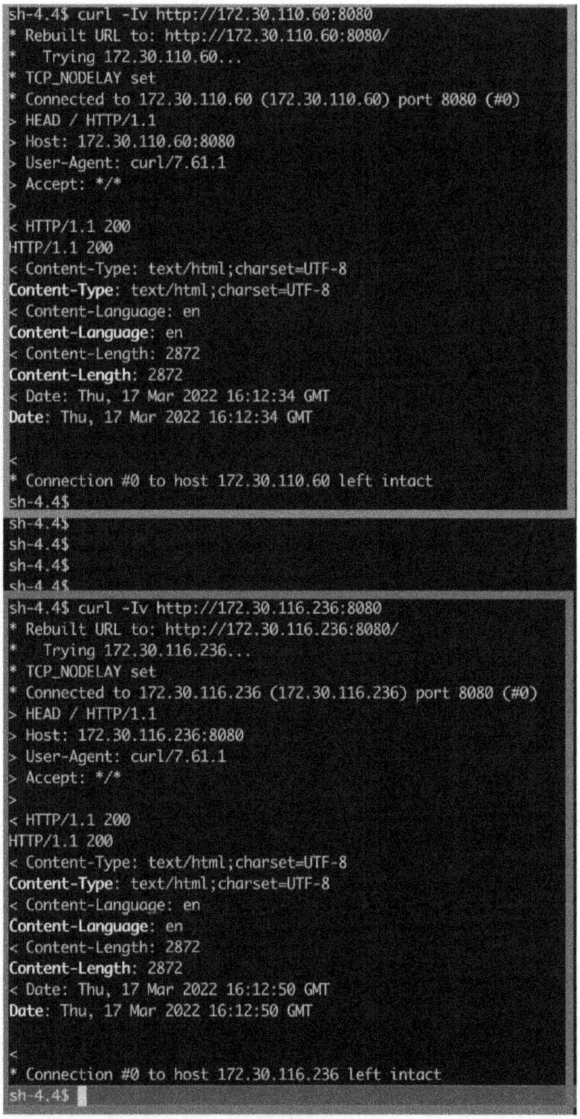

Figure 7.7 – Testing connectivity between two namespaces

In our next scenario, we will block all traffic on the greenpets namespace, for which the diagram looks like the following:

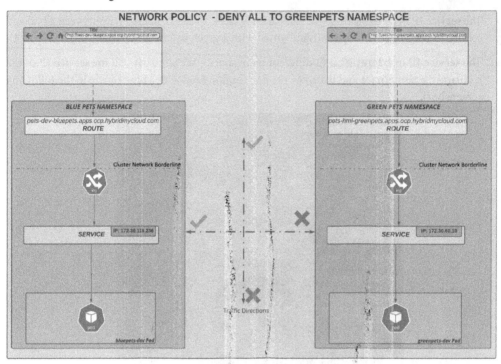

Figure 7.8 – greenpets namespace – denying all traffic

To accomplish this scenario, we apply a network policy manifest on the greenpets namespace:

```
$ cat << EOF >> block-everything-to-namespace.yaml
kind: NetworkPolicy
apiVersion: networking.k8s.io/v1
metadata:
  name: deny-by-default
spec:
  podSelector:
  ingress: []
EOF
$ oc -n greenpets apply -f block-everything-to-namespace.yaml
```

Now, let's perform the same tests again to demonstrate that all network traffic in greenpets (route and service) is denying connections:

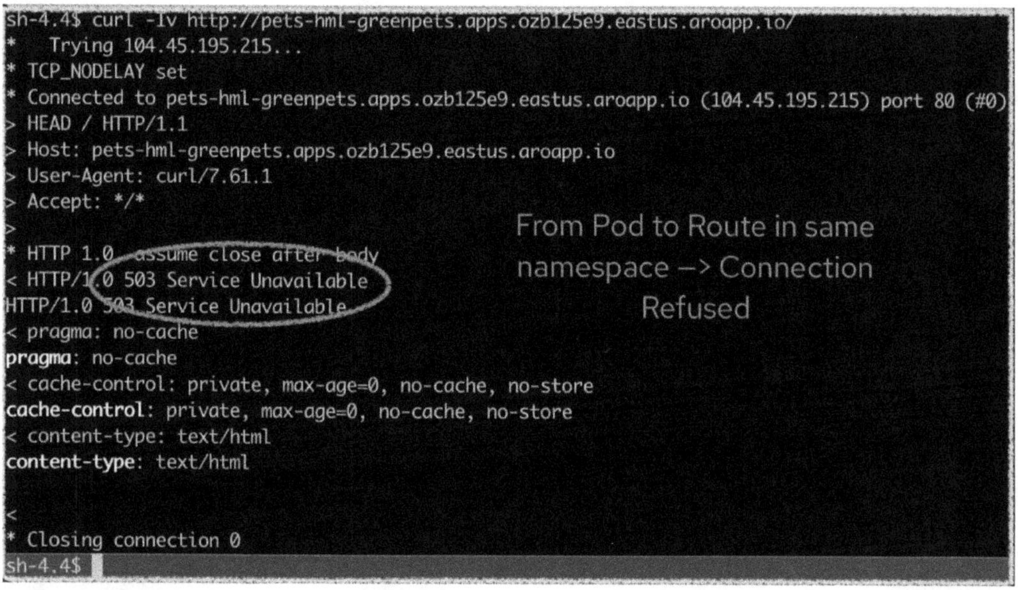

Figure 7.9 – Deny all traffic test

Now, we will go deeper and apply a rule that only allows traffic from ingress to flow to pods under the `greenpets` namespace. To do so, we are going to apply the following YAML file:

```
$ cat << EOF >>allow-ingress-to-namespace.yaml
apiVersion: networking.k8s.io/v1
kind: NetworkPolicy
metadata:
  name: allow-from-openshift-ingress
spec:
  ingress:
  - from:
    - namespaceSelector:
        matchLabels:
          network.openshift.io/policy-group: ingress
  podSelector: {}
  policyTypes:
  - Ingress
EOF
$ oc -n greenpets  apply -f allow-ingress-to-namespace.yaml
```

What this NP does is to only allow pods in the ingress namespace to communicate with pods in the greenpets namespace, all other traffic will be blocked. Check out the following diagram and notice that *east-west* traffic between namespaces is denied, but *north-south* traffic is allowed:

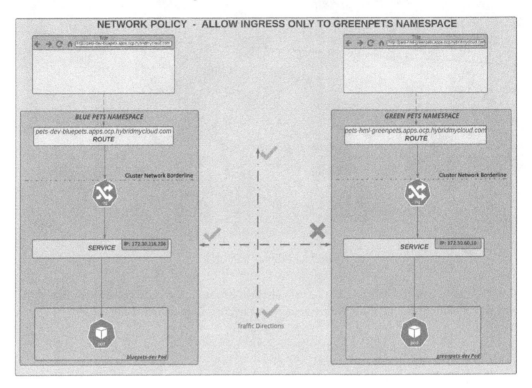

Figure 7.10 – greenpets namespace traffic only allowed for Ingress connections (external route)

Notice now that the network communication between the external route (ingress) and the service is working; however, traffic between bluepets and greenpets is denied.

```
% oc -n bluepets rsh pets-dev-7846d7d75-sdkc8
sh-4.4$
sh-4.4$
sh-4.4$ curl -Iv http://172.30.110.60:8080                    1
* Rebuilt URL to: http://172.30.110.60:8080/
*   Trying 172.30.110.60...
* TCP_NODELAY set
* connect to 172.30.110.60 port 8080 failed: Connection timed out
* Failed to connect to 172.30.110.60 port 8080: Connection timed out
* Closing connection 0
curl: (7) Failed to connect to 172.30.110.60 port 8080: Connection timed out
sh-4.4$ curl -Iv http://pets-hml-greenpets.apps.
*   Trying 104.45.195.215...
* TCP_NODELAY set
* Connected to pets-hml-greenpets.apps.                    (104.45.195.215) port 80 (#0)
> HEAD / HTTP/1.1
> Host: pets-hml-greenpets.apps.
> User-Agent: curl/7.61.1
> Accept: */*
>
< HTTP/1.1 200
HTTP/1.1 200
< content-type: text/html;charset=UTF-8                    2
content-type: text/html;charset=UTF-8
< content-language: en
content-language: en
< content-length: 2872
content-length: 2872
< date: Thu, 17 Mar 2022 19:31:04 GMT
date: Thu, 17 Mar 2022 19:31:04 GMT
< set-cookie: 78fbf668d708c59275ac7bedf6c54540=d77151b1491f732de74affaec80db50d; path=/; HttpOnly
set-cookie: 78fbf668d708c59275ac7bedf6c54540=d77151b1491f732de74affaec80db50d; path=/; HttpOnly
< cache-control: private
cache-control: private
<
* Connection #0 to host pets-hml-greenpets.apps.                    left intact
sh-4.4$
```

Figure 7.11 – Testing network traffic. 1) From bluepets namespace to greenpets namespace:
Connection denied. 2) From external route (ingress) to greenpets namespace: Connection allowed.

Finally, we will take a look at the most common scenario: the least isolation configuration. This network policy scenario is based on a namespace label that we will apply in the greenpets namespace and will work as a key to configure the communication between namespaces.

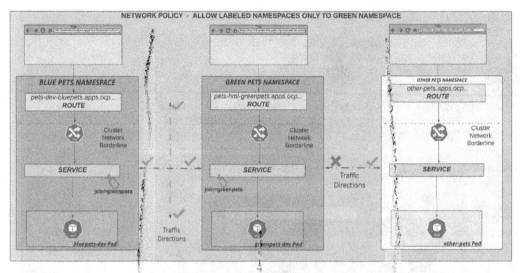

Figure 7.12 – Labeled namespaces allowing traffic

Looking at the previous diagram, you can see three different namespaces, `bluepets`, `greenpets`, and `otherpets`. A network policy will be applied to the `greenpets` namespace, which will use a label with the `join=greenpets` value. In other words, it means that only elements in namespaces labeled with `join=greenpets` can communicate with the application in the `greenpets` namespace. To implement this, we will apply the following manifest and commands:

```
$ cat << EOF >> allow-namespace-by-label.yaml
apiVersion: networking.k8s.io/v1
kind: NetworkPolicy
metadata:
  name: allow-from-namespace-label
spec:
  ingress:
  - from:
    - namespaceSelector:
        matchLabels:
          join: greenpets
  podSelector: {}
EOF
$ oc -n greenpets apply -f allow-namespace-by-label.yaml
$ oc label namespace bluepets join=greenpets
```

Now, check the connectivity between the namespaces `bluepets` and `greenpets` by running the following test:

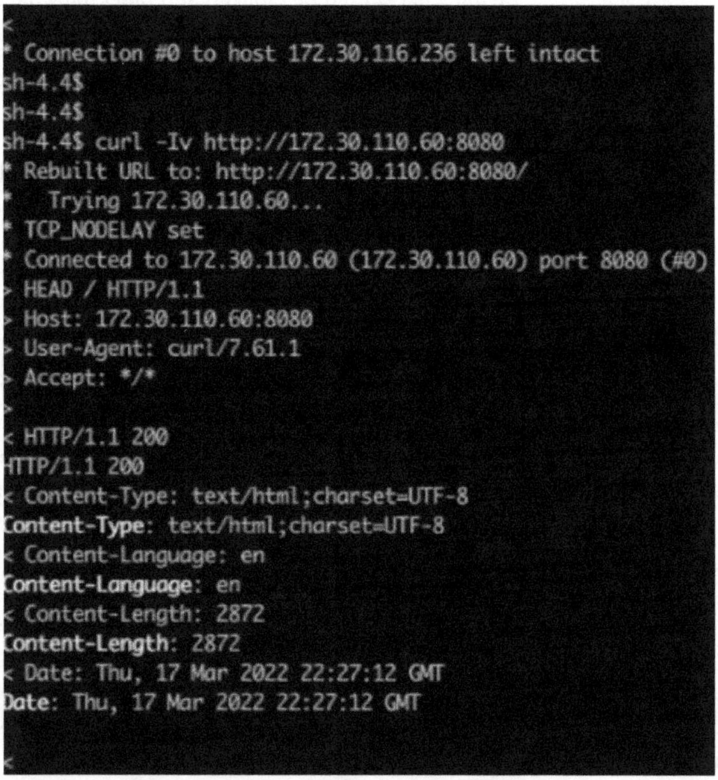

```
<
* Connection #0 to host 172.30.116.236 left intact
sh-4.4$
sh-4.4$
sh-4.4$ curl -Iv http://172.30.110.60:8080
* Rebuilt URL to: http://172.30.110.60:8080/
*   Trying 172.30.110.60...
* TCP_NODELAY set
* Connected to 172.30.110.60 (172.30.110.60) port 8080 (#0)
> HEAD / HTTP/1.1
> Host: 172.30.110.60:8080
> User-Agent: curl/7.61.1
> Accept: */*
>
< HTTP/1.1 200
HTTP/1.1 200
< Content-Type: text/html;charset=UTF-8
Content-Type: text/html;charset=UTF-8
< Content-Language: en
Content-Language: en
< Content-Length: 2872
Content-Length: 2872
< Date: Thu, 17 Mar 2022 22:27:12 GMT
Date: Thu, 17 Mar 2022 22:27:12 GMT
<
```

Figure 7.13 – Testing labeled namespace. Connection to a namespace
that contains the proper label – connection allowed

In *Figure 7.13*, you see that the connection was allowed as the namespace contains the label `join=greenpets`. However, in *Figure 7.14*, you can see the connection is denied, as the traffic flows from a namespace (`otherpets`) that doesn't contain this label.

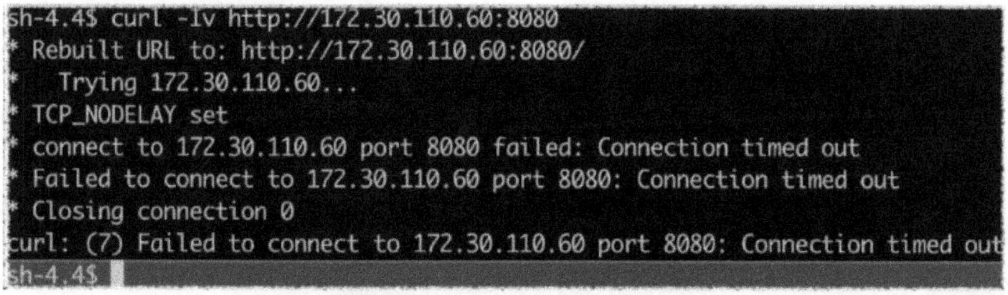

```
sh-4.4$ curl -Iv http://172.30.110.60:8080
* Rebuilt URL to: http://172.30.110.60:8080/
*   Trying 172.30.110.60...
* TCP_NODELAY set
* connect to 172.30.110.60 port 8080 failed: Connection timed out
* Failed to connect to 172.30.110.60 port 8080: Connection timed out
* Closing connection 0
curl: (7) Failed to connect to 172.30.110.60 port 8080: Connection timed out
sh-4.4$
```

Figure 7.14 – Testing non-labeled namespace denying traffic

Network policy is an important tool to isolate network traffic. It is important you consider the challenges that certain types of rules may bring, though. If not properly designed, standardized, and adopted, they may cause you headaches by allowing what should be blocked and blocking what shouldn't be.

Also, you have to consider which types of workload will run in your cluster. For microservice-oriented applications, for instance, we recommend you look at the **Istio service mesh**, which in general is more appropriate and will bring more granular network access control.

So far, you have learned the definitions and important concepts of SDNs, such as controlling traffic in horizontal and vertical directions by applying policies using labels. Continue, next, to see more about routes and ingress controllers and learn how to use them for your applications.

What is an ingress controller?

An **Ingress controller** is a lightweight, self-healing load balancer that distributes network traffic from outside the cluster to a network service. Using an Ingress controller is a standard approach for providing and managing ingress traffic to containerized applications. The default ingress controllers on OpenShift use the mature and stable **HAProxy** under the hood. In OpenShift, when you deploy a cluster, the ingress controller is automatically created and hosted in two worker nodes by default.

How does an ingress operator work?

An Ingress operator acts similarly to almost all cluster operators in OpenShift: protecting the important settings of the operation of a cluster. The operator monitors the ingress pods running in the `openshift-ingress` namespace and protects the `IngressController` objects from wrong and non-compatible settings that can lead to problems with the cluster network.

Otherwise, you can create others `IngressController` objects in addition to the default one to isolate the traffic of certain groups of applications, using what is named **router sharding**.

Different from traditional networking configuration, in which you need complex routing tables and firewall configuration, OpenShift abstracts this complex networking layer configuration, making it a much easier task.

Creating a new ingress controller

To create a new ingress controller, you must take the following steps:

1. Define at least two nodes to host the new ingress controller.

2. Apply a new label to nodes.

3. Export the default `IngressController` object.

4. Change the name and desired settings of the newly created YAML manifest file.

5. Deploy the new `IngressController` object by applying the YAML created previously.

You can see in the following lines an example of the process mentioned previously:

```
# Apply labels to nodes
$ oc label node <node1> <node2> .. <nodeN> new-ingress=true
$ oc get ingresscontroller default -n openshift-ingress-
operator -o yaml > new-ingress.yaml
$ vi new-ingress.yaml
# Remove unnecessary fields to make the yaml looks like
# the following one
apiVersion: operator.openshift.io/v1
kind: IngressController
metadata:
  name: new-ingress [1]
  namespace: openshift-ingress-operator
spec:
  domain: apps.env.hybridmycloud.com [2]
  replicas: 2
  nodePlacement:
    nodeSelector:
      matchLabels:
        new-ingress: "true" [3]
  routeSelector: [4]
    matchLabels:
      type: sharded [5]
```

In the previous code, we have highlighted some parts with numbers. Let's take a look:

[1]: New IngressController name.

[2]: DNS domain for the new ingress.

[3]: A label that defines where the IngressController pods will run.

[4]: To implement shards. It can be namespaceSelector or routeSelector.

[5]: Used to filter the set of routes that are served by this IngressController.

Namespace or Route Selector?

The example you have seen uses the `routeSelector`. There is an alternative way to configure the IngressController, which is using `namespaceSelector`. It may seem confusing to define the right selector for your case, but it is not – `routeSelector` is a more granular option, allowing you to publish routes to different IngressControllers in the same namespace. The main decision factor is if, in your case, you need to be able to publish routes of a single namespace in different IngressControllers, you have to use `routeSelectors`. Otherwise, you will most likely use `namespaceSelectors`.

For example, consider a namespace called `APP` that contains two different routes:

Route A published in router 1 with the URL `app1.prod.hybridmycloud.com`

Route B published in router 2 with the URL `app1.qa.hybridmycloud.com`

This scenario is only possible if you use `routeSelector`. However, this is an unusual scenario; usually, routes in a single namespace are always published in the same IngressController, so for that reason, it is also very common to use `namespaceSelector`.

As previously mentioned, router sharding is a technique that allows creating an ingress for the purpose of segregating traffic, whether due to the need for isolation between environments or even for the traffic of a given application to be fully directed from this new ingress.

Testing the new ingress

After the ingress pods are created on the nodes, you can test the newly created ingress. We will create a route using the sample application named `hello-openshift` and apply the proper route selector label. Follow these steps to accomplish this task:

```
$ oc new-project hello-openshift
$ oc create -f https://raw.githubusercontent.com/openshift/
origin/master/examples/hello-openshift/hello-pod.json
```

```
$ oc expose pod/hello-openshift
$ oc expose svc hello-openshift
$ oc label route hello-openshift type=sharded
```

The last line of the previous block of commands explicitly sets the `type=sharded` label, which we used in our example for `routeSelector`. When OpenShift sees this label, it will automatically publish this route in the new ingress.

Continue on to the following section to get a full understanding of how to use the recently created ingress with what is called a **route** in OpenShift.

Types of routes

Routes are the representation of a configuration on an ingress internal load balancer for a specific application to expose a Kubernetes service to a DNS name, such as `example.apps.env.hybridmycloud.com`. When a route is created, OpenShift automatically configures a frontend and backend in the Ingress' HAProxy pod to publish the URL and make the traffic available from the outside world.

Routes can be published using either the HTTP or HTTPS protocol. For HTTPS, three different types of routes define how the TLS termination works in the SSL stream between the user and the pod. In the following subsections, we will walk you through each of them.

Passthrough routes

A **passthrough route**, as the name suggests, is a configuration in which the packages are forwarded straight to the network service without doing a TLS termination, acting as a Layer 4 load balancer. Passthrough is often used with applications that provide their own TLS termination inside the application's pod, either by implementing it in the source code or using a middleware layer (such as JBoss or WebSphere).

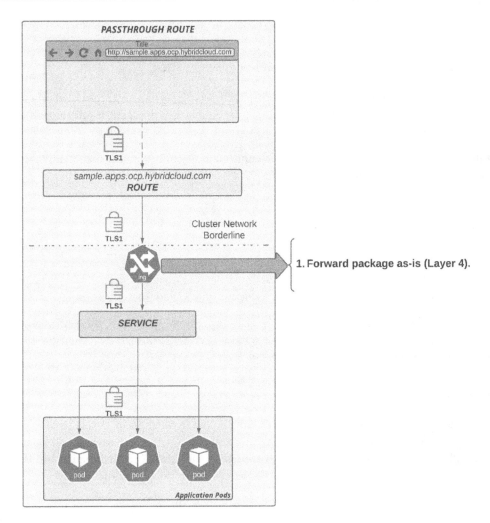

Figure 7.15 – Passthrough route

Next, you'll see the second option you have: edge route.

Edge routes

In this route, the TLS termination is handled by OpenShift ingress and forwarded to the service as clear text. This kind of route is used very often as it is easy to use: a self-signed certificate automatically generated by OpenShift is applied to the ingress and it signs all the routes that use the default wildcard domain – this is performed by OpenShift automatically; no additional configuration is needed. However, you can replace the self-signed certificate with a custom digital certificate, if you don't want to use the default self-signed certificate.

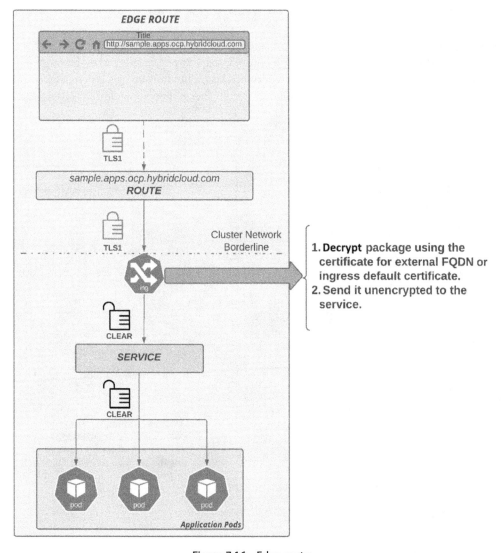

Figure 7.16 – Edge route

An edge route is the most common and easy-to-implement model since the certificate chain terminates at the edge of the OpenShift network, which is the ingress. It is important to highlight that the traffic between the ingress and the application pods is not encrypted but occurs inside the OpenShift SDN, which means that the network packages are encapsulated using OVS. The last method available is reencrypted routes. You'll see how it works next.

Reencrypted routes

Reencrypted routes offer two layers of TLS termination: traffic is decrypted using the certificate for the external FQDN (for example, `example.apps.env.hybridmycloud.com`) at the cluster edge (OpenShift Ingress), and then the traffic is re-encrypted again, but now using a different certificate. While this is a secure route, it has also a performance penalty due to the termination and re-encryption operation performed by the ingress.

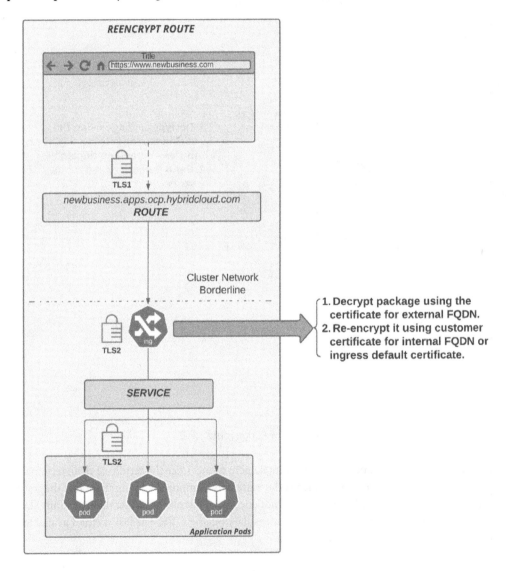

Figure 7.17 – Reencrypted route

A reencrypted route takes a similar approach as an edge route but it goes through two layers of CAs. The first is related to the external public domain, for example, *hybridcloud.com*, and then the second layer of encryption is internal, known by OpenShift Ingress and the application.

Summary

We have seen in this chapter some of the important aspects related to the OpenShift network. Now you are familiar with the two types of network plugins supported with OpenShift, OpenShift SDN and OVN-Kubernetes, and the different kinds of traffic you need to care about when managing the platform's network. You have also seen how the ingress controller works, how to create a new one, and the three different types of secure routes you may use with your applications: passthrough, edge, and reencrypted.

You navigated through network policies to learn a bit more about how to control traffic and provide network isolation.

As you know, security is a real concern in today's digital world. In the next chapter, we will cover important aspects you need to consider about security on OpenShift. So, go ahead and check it out!

Further reading

If you want more information related to the concepts we covered in this chapter, check out the following references:

- *Kubernetes Ingress controller*: `https://www.nginx.com/resources/glossary/kubernetes-ingress-controller`

- *HAProxy documentation*: `https://www.haproxy.com/documentation/hapee/latest/onepage/`

- *Annotations used to override a route's default configuration*: `https://docs.openshift.com/container-platform/4.10/networking/routes/route-configuration.html#nw-route-specific-annotations_route-configuration`

- *Configuring ingress cluster traffic using an Ingress controller*: `https://docs.openshift.com/container-platform/4.10/networking/configuring_ingress_cluster_traffic/configuring-ingress-cluster-traffic-ingress-controller.html`

- *Creating secured routes*: `https://docs.openshift.com/container-platform/4.10/networking/routes/secured-routes.html`

- *OpenShift Tested Integrations*: `https://access.redhat.com/articles/4128421`

- *Service mesh*: `https://docs.openshift.com/container-platform/4.10/service_mesh/v2x/servicemesh-release-notes.html`

8

OpenShift Security

Throughout this book, you have been able to follow some important aspects that involve processes, people, and all the technology involved to maintain a powerful, resilient, and fault-tolerant platform. A product of this magnitude with a vast structure, from its creation to customizations that will keep it fully functional, also requires a great division of responsibilities and skills for each step taken.

Security has always been and will always be a top concern in any enterprise, even more so now with the advent of things such as ransomware, malicious crypto-mining, malware, and other types of attacks. According to a *Gartner* study, by 2025, 90% of organizations that don't control public cloud usage properly will inadvertently share sensitive data. When it comes to Kubernetes security, we also have seen a number of vulnerabilities and attacks recently. In the *Further reading* section of this chapter, you can check out some vulnerabilities and security incidents related to Kubernetes that happened in the last few years.

For this reason, we decided to dedicate an entire chapter to security-related aspects involving OpenShift. The main focus of this chapter is not to be a complete security guide, though, but rather a minimal collection of topics you should consider and review to have a secure container strategy.

In this chapter, you will find the following topics:

- Container security
- AuthN and AuthZ
- Certificates in OpenShift
- etcd encryption
- Container isolation
- Network isolation
- Red Hat Container Catalog

Let's get started!

Container security

Container adoption grows exponentially, and with that also the number of vulnerabilities and potential attacks against all the container ecosystem. That being said, security is an important aspect you need to consider in your container strategy. Red Hat is well known for the high level of security of its products, which is one of the characteristics that has always been a differentiator for them in the industry, since the robust **Red Hat Enterprise Linux** (**RHEL**), which was (and still is!) the foundation of the company up to emerging technologies, such as Red Hat OpenShift. They have been named leaders in the container security space due to a number of security features that Red Hat defined as default (and required) with OpenShift that are optional in many other Kubernetes-based platforms, which also makes OpenShift more secure than other options. One example of it is **Security-Enhanced Linux** (**SELinux**), which is always enabled in any OpenShift worker nodes and prevents a number of vulnerabilities and exploits.

> **Did you know?**
>
> Did you know that there are now probably somewhere about 380,000 Kubernetes **application programming interfaces** (**APIs**) open on the internet for some form of access that are probably exposed to some sort of attack or data leak? That is what *The Shadowserver Foundation* found in this research: `https://www.shadowserver.org/news/over-380-000-open-kubernetes-api-servers/`.
>
> Interested in seeing more? In this *Container Security* report, you will find some great research related to security on Kubernetes: `https://www.kuppingercole.com/reprints/b1e948f62d5394353f996e43a89cde4a#heading8.1`.
>
> Here, you can view the *State of Kubernetes security report*: `https://www.redhat.com/en/resources/kubernetes-adoption-security-market-trends-overview`.
>
> You can also consult the *Is the Cloud Secure? Gartner* study at the following link: `https://www.gartner.com/smarterwithgartner/is-the-cloud-secure`.

The key to having a secure environment and applications resides in the ability that a company has to do the following:

- **Control**: Secure the **application life cycle management** (**ALM**) to detect and fix vulnerabilities before it goes live

- **Protect**: Ability to assess and protect the platform and infrastructure to avoid vulnerabilities being exploited

- **Detect and respond**: Detect and mitigate vulnerabilities by limiting their impact on systems and environments

In the following diagram, you will see some aspects of these three factors we mentioned:

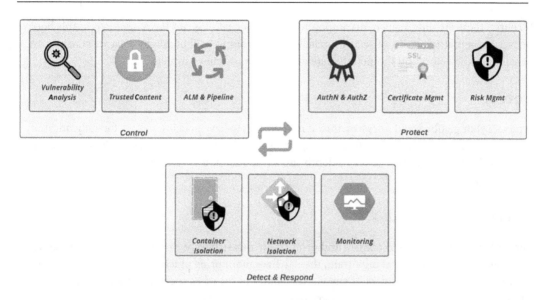

Figure 8.1 – Container security strategy

In this chapter, we will walk through some important things to consider about each of these factors.

Control

To secure the container software supply chain, a few common steps can be performed, as follows:

- **Vulnerability analysis**: Scan the container images to find known vulnerabilities. This topic will be covered in detail in *Chapter 12, OpenShift Multi-Cluster Security*.

- **Trusted images and contents**: Do you know everything that is running inside your containers and applications? Every application these days uses many dependencies from different projects, many of them open source. It is crucial to use trusted sources to avoid running compromised base packages and images. Red Hat provides the Red Hat **Universal Base Image** (**UBI**), which is the robust, secure, and stable version of RHEL that you can use as a base for your container images. Red Hat also provides a comprehensive list of certified container images through its Red Hat Container Catalog that are safe to use and maintains a Container Health Index that helps to assess the security risk of a container image.

- **ALM and continuous integration/continuous deployment (CI/CD) pipeline**: Security should be an integral part of your development workflow. A modern development life cycle requires that security be a shared responsibility among all teams, from end to end. **Development-security-operations** (**DevSecOps**) is the new norm, and that means automating security gates to integrate security into the **development-operations** (**DevOps**) processes. Things such as **integrated development environment** (**IDE**) plugins and CI/CD pipeline security tasks are must-haves to achieve this.

Of course, these are only a few items among a much bigger set. It is not the main focus of this book to cover a comprehensive list of security-related aspects, but we will highlight what we consider are some important factors you should take care of with your container strategy.

Protect

To protect your platform and infrastructure, it is important to have policies in place for authentication and authorization, deployment, certificate management, compliance assessment, and policy enforcement. Later in this chapter, we will cover in depth the authentication and authorization aspects of an OpenShift cluster.

Detect and respond

Even with all actions taken to control and protect the infrastructure, you need to be able to detect and mitigate security risks. It is important, then, to have monitoring practices in place, but also other policies that may limit the impact of any vulnerabilities or breaches such as container and network isolation. In this chapter, you will also see some more information about container and network isolation.

Let's start by looking at authentication and authorization in more detail. Continue reading to learn more about them.

AuthN and AuthZ

AuthN and **AuthZ** are related concepts but with different goals. **AuthN** is an abbreviation of **auth**entication and is related to the process to confirm who a user is and validate their credentials to grant access to the platform. In this process, we deal with **identity providers (IdPs)** on OpenShift to configure the authentication provider that will check the user credentials, which we will cover shortly.

In turn, **AuthZ** stands for **auth**orization and is directly linked to giving the user only what they really should have access to, after the authentication process.

These two concepts are seen as confusing or similar from the point of view of just logging in to OpenShift. To clarify the concepts, we will now discuss the different objects and processes related to authentication and authorization.

Authentication

OpenShift works with the following set of objects as part of the authentication process:

Object	Description
Users	Resource represented by a human figure.
Groups	This object represents a group of users that inherit common permissions.
IdP	Authentication provider that OpenShift relies on to validate a user and give access (or not) to the platform. You will find a list of supported IdPs later in this chapter.
Service account	A service account allows you to control API access without using a regular user. It is very useful for application and external tool authentication through OpenShift APIs.

These are the objects used in OpenShift that, combined, authenticate a user in the platform. In order to understand how authentication works, you need to define IdPs on OpenShift—see more information about them next.

IdPs

IdPs are entities that integrate OpenShift with authentication providers, with the purpose of creating a chain of custody over who can log on to the platform. To avoid the privilege of one product over another when it comes to managing users, OpenShift creates a very flexible option that allows multiple IdPs to integrate with OpenShift at the same time.

The following table contains a list of supported IdPs allowed for OpenShift integration:

Provider	Description
htpasswd	Apache project's open source authentication provider. Recommended as a second method to keep cluster authenticating when the first authentication provider is unavailable for some reason.
Lightweight Directory Access Protocol (LDAP)	LDAP is a large-scale adopted form of authentication that uses the Berkeley Internet Name Domain (BIND) protocol. It can be found on an OpenLDAP implementation such as Identity Manager (IdM) and also in proprietary authentication software such as Microsoft Active Directory (AD). Both support a secure way of authenticating, using digital certificates to encrypt user data.
GitHub/GitLab	Source code versioning platforms based on GitHub or GitLab are allowed to integrate the user into OpenShift clusters.
OpenID Connect (OIDC)	Use an OIDC provider to authenticate users using the Open Authorization (OAuth) protocol.
Request header	A request header IdP is used when an authentication proxy is required to identify users from request header values, such as the X-Remote-User field, which sets the request header value.
Basic authentication	This is an IdP that permits external providers to authenticate. Users send their username and password to OpenShift Container Platform, which then validates those credentials, making a server-to-server authentication. It is useful for enterprises that have developed their own authentication provider.
Keystone	This is the default IdP used on Red Hat OpenStack that can be used to authenticate users on OpenShift when a customer is already using this platform.

In the OpenShift documentation, you will find detailed information about how to configure each of these providers: `https://docs.openshift.com/container-platform/latest/authentication/index.html`.

OpenShift uses a **role-based access control** (**RBAC**) model to perform the authorization process and give a user the appropriate set of permissions they should have, as you will see now.

Authorization – RBAC

In an RBAC system, you will have a set of permissions represented as a *role* that users and groups will use to have those permissions granted. On OpenShift, we have the following objects to configure RBAC permissions:

Object	Description
Rules	Rules rely on verbs and objects to set up permissions. For instance, a user can list (verb) secrets (object) or create (verb) pods (object).
Roles	Roles are a collection of rules that can be associated with users and/or groups—for example, the View role, which contains all rules that grant read-only access to see most objects in a project.
Bindings	A binding is an association between roles and users or groups—for example, monitoring-users associates the monitoring group with the role that contains all the rules/permissions for application monitoring.

A role can be either cluster- or local-scoped, which means that it can be applied for all projects or only for a specific project. In the following example, you can see a command to create a *local* role that will allow a user to get and update a pod in the namespace test:

```
$ oc create role podmgmt --verb=get,update --resource=pod -n
test
```

You can check the role created by running the following command:

```
$ oc describe role.rbac.authorization.k8s.io/podmgmt -n test
Name:         podmgmt
Labels:       <none>
Annotations:  <none>
PolicyRule:
  Resources  Non-Resource URLs  Resource Names  Verbs
  ---------  -----------------  --------------  -----
  pods       []                 []              [get update]
```

Similarly, to create a *cluster* role, you need to use the `oc create role` command, like so:

```
$ oc create role clusterpodmgmt --verb=get,update
--resource=pod
```

To create a role binding to add this role to a user or group, the following commands can be used:

```
# For local role bindings
$ oc adm policy add-role-to-user <role> <username> -n <project>
$ oc adm policy add-role-to-group <role> <group> -n <project>
# For cluster role bindings
$ oc adm policy add-cluster-role-to-user <role> <username>
$ oc adm policy add-cluster-role-to-group <role> <group>
```

You can also remove a role from a user or group using the following commands:

```
# For local role bindings
$ oc adm policy remove-role-to-user <role> <username> -n
<project>
$ oc adm policy remove-role-to-group <role> <group> -n
<project>
# For cluster role bindings
$ oc adm policy remove-cluster-role-to-user <role> <username>
$ oc adm policy remove-cluster-role-to-group <role> <group>
```

Continue reading for hints to troubleshoot issues you may have related to the access-control feature.

Access-control troubleshooting

It is important you notice that after you change the cluster OAuth custom resource, the authentication operator will apply the change; due to that, you may observe temporary unavailability of the OpenShift authentication. Allow some minutes for the operator to finish applying the changes.

Even if you face issues with OpenShift authentication, we recommend you check out this helpful article: `https://access.redhat.com/articles/5900841`.

Certificates in OpenShift

OpenShift uses several different certificates to make the platform secure. In this section, we will walk through the main certificate chains OpenShift uses. These are presented here:

- **API server**: Certificate used with the OpenShift API, usually accessible at `https://api.<cluster-name>.<domain>:6443/`.

- **Ingress**: Certificate valid for application's domain. It is usually a wildcard certificate for `*.apps.<cluster-name>.<domain>`.

- **Node certificates**: Managed automatically by the cluster and don't require any user intervention. Automatically rotated.

- **etcd certificates**: Used to encrypt communication between etcd cluster members. Automatically managed and rotated by the cluster itself.

OpenShift creates self-signed certificates with the platform installation. Most enterprises require those certificates to be replaced by custom certificates. To do so, refer to the OpenShift documentation, as follows:

- To replace the ingress certificate: `https://docs.openshift.com/container-platform/latest/security/certificates/replacing-default-ingress-certificate.html`

- To replace the API server: `https://docs.openshift.com/container-platform/4.10/security/certificates/api-server.html`

If you replace the certificates, you may also need to add a trusted **certificate authority** (**CA**), if you use a private CA. The next section describes how to do this.

Trusted CA

You may need to add a custom CA that you use within your organization, to make API calls possible between OpenShift components and other external systems. To do so, you need to create a config map with the certificate chain on the `openshift-config` namespace, like so:

```
$ oc create configmap custom-ca \
      --from-file=ca-bundle.crt=</path/to/example-ca.crt> \
      -n openshift-config
```

Then, update the cluster-wide proxy to use the config map just created, as follows:

```
$ oc patch proxy/cluster \
    --type=merge \
    --patch='{"spec":{"trustedCA":{"name":"custom-ca"}}}'
```

One important piece of any Kubernetes cluster is the etcd database. You may decide to encrypt it to keep secrets and other objects encrypted in the database. Check out next how to do this.

etcd encryption

The etcd database, by default, is not encrypted. However, you can easily enable etcd encryption to have an extra layer of data security in your cluster. When etcd encryption is enabled, things such as secrets and config maps are stored encrypted, which makes your cluster even more secure.

To enable etcd encryption, proceed as follows:

1. Edit the `apiserver` object, like so:

    ```
    $ oc edit apiserver
    ```

2. Set the `encryption` field, as follows:

    ```
    spec:
      encryption:
        type: aescbc
    ```

3. Save the file to apply the changes.

Note that it may take up to 15 minutes or so to enable the encryption after you have applied the changes.

Container isolation

We already discussed aspects such as user authentication and permissions, as well as certificates, but how do you make sure your containers can only do what they are supposed to do and nothing more, and as such, cannot escalate privileges on the host?

We are going to discuss in this section some of the concepts implemented as part of the **operating system (OS)** to securely run containers and also some aspects implemented on OpenShift related to this.

In the following table, you see some concepts related to containers:

Concept	Description
Linux namespaces	Namespaces are a Linux kernel feature that provides an isolation level for processes and resources.
Control groups (Cgroups)	Limiting access to system resources. Through Cgroups, it is possible to limit the amount of central processing unit (CPU), memory, and other OS resources a container has.
SELinux	SELinux is a security layer that aims to protect user data on Linux OSs. In an SELinux-enabled system, every file, process, directory, and port has a label called an SELinux context, and a process can only access resources that SELinux rules allow it to. On OpenShift, all container processes have a `container_t` context label, while files and directories that the container use have a `container_file_t` context. Additionally, OpenShift adds SELinux rules to allow processes in the `container_t` domain to only access `container_file_t` files. Therefore, if a process in a container tries to escape its own space and access other files on the host system, the SELinux feature will block the access and prevent any attempt to access the files. SELinux is always enabled in any worker node on OpenShift, and that itself prevents several security issues detected on Kubernetes from being exploited on OpenShift.

In the following diagram, you can see a graphical representation of these concepts:

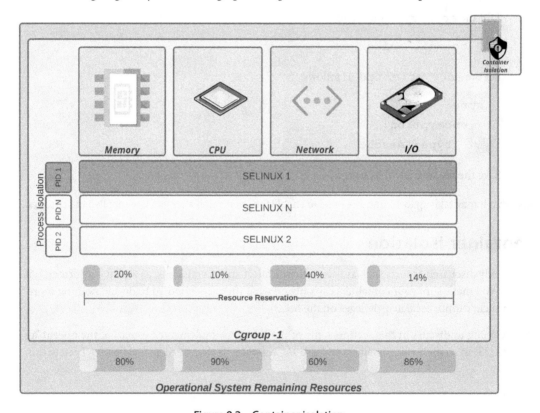

Figure 8.2 – Container isolation

While it is important to understand these concepts, you don't really need to work directly with any of them, as OpenShift abstracts those configurations for you. Instead, you need to understand what **security context constraints (SCCs)** are—these are the objects you will need to use to give broader permissions to containers when needed. Read on to find out what SCCs are and how to configure them.

SCCs

SCCs are OpenShift objects that control the actions that a pod or container can do with the host resources. By default, any pod uses the **restricted** SCC, which is the most restricted permission that prevents it from running as root and escalating privileges on the host. Besides the restricted SCC, the most usual ones are noted here:

- `anyuid`: This has the same permissions as the restricted SCC, allowing a container to run as any **unique identifier (UID)** and **group ID (GID)**, though that means the container can also use the 0 UID—the root user.

- `hostaccess`: Gives permissions for a container to access all host namespaces. It should be used only with trusted sources as it gives very broad permissions within a worker node—use only if really necessary.

- `hostnetwork`: Allows a container to access the worker node underlying the network directly. Use with caution to decrease the risk of a network breach.

- `privileged`: Most relaxed SCC and gives all privileges to the host. Should be avoided at all costs and used only if really necessary.

Besides all the security-related features we already covered so far, we cannot miss a very important topic: network policies. A network policy is a great feature to make sure your pods and projects can only communicate with what they are supposed to. In the following section, you will see what you can do with network policies and why they are so important when it comes to network security.

Network isolation

Firewalls are well known and have been used for a long time in any kind of infrastructure. When it comes to OpenShift, we need to have in mind that we are now working with a software-defined platform and, as such, we have software features to implement some of the same concepts we have had for a long time in a data center—it is no different with a firewall. As we have seen in the previous chapter, Network Policies are nothing more than rules you define to allow or block network communication between pods and projects on OpenShift, similar to what a firewall provides in a physical network.

By default, all pods in a project are accessible from other pods and network endpoints from any project. To isolate pods and projects, you need to create network policies, such as the following:

```
kind: NetworkPolicy
apiVersion: networking.k8s.io/v1
```

```
metadata:
  name: deny-by-default
spec:
  podSelector: {}
  ingress: []
```

The previous network policy denies all traffic for pods in a project. In order for a project to be accessed from outside the cluster using the OpenShift ingress, you will need to allow connections from the ingress project, as you can see in the following snippet:

```
apiVersion: networking.k8s.io/v1
kind: NetworkPolicy
metadata:
  name: allow-from-openshift-ingress
spec:
  ingress:
  - from:
    - namespaceSelector:
        matchLabels:
          network.openshift.io/policy-group: ingress
  podSelector: {}
  policyTypes:
  - Ingress
```

You may also create custom network policies using labels for specific pods. For instance, the following YAML Ain't Markup Language (YAML) could be used to limit pods with an app=web label to be accessed on the accessed on the HTTPS port (443) only port only:

```
kind: NetworkPolicy
apiVersion: networking.k8s.io/v1
metadata:
  name: allow-https
spec:
  podSelector:
    matchLabels:
      app: web
  ingress:
  - ports:
```

```
  - protocol: TCP
    port: 443
```

There is a world of possibilities for using network policies to set your projects and pods with proper network access. A common and recommended practice is to create a set of standard network policies and configure the OpenShift template so that those network policies will be automatically applied to any new projects created. You can find out how to customize the OpenShift template to add those network policies in the OpenShift documentation: `https://docs.openshift.com/container-platform/latest/networking/network_policy/default-network-policy.html`.

Another important thing you may consider in your security strategy is about using safe container base images. The following section covers what the Red Hat Container Catalog is and how it can help you with this important topic.

Red Hat Container Catalog

Most enterprises, at some point, need to use containers from different sources, either as off-the-shelf products or only as dependencies for workloads. Red Hat has an interesting tool that aims to help you to make sure you use secure containers. The Red Hat Container Catalog is a repository of container images that have been tested and certified by Red Hat and partners, and therefore are trusted images.

As part of the Container Catalog, Red Hat provides the Container Health Index. The Container Health Index is a rating system for container images available in the Red Hat Container Catalog, which goes from Grade A to F, in which Grade A is the most up-to-date image in terms of critical errata. You can find all the grades available and how they are defined at this link: `https://access.redhat.com/articles/2803031`. This feature helps a lot to assess the security risk you are associated with when using a certain image version.

In the Red Hat Container Catalog, you can find the Red Hat UBI, which may comprise an important aspect of your container strategy. You'll find out more about this next.

Red Hat UBI

Defining a base image for your containers is an important step that you need to consider to have a portable and secure container delivery process. Depending on the decision you make, you may see yourself locked in some specific Linux versions and distributions that the container base image supports. The base image is also critical to keep your workloads secure—many vulnerabilities are found every week on Linux, but they are usually fixed really fast, using a mature and robust base image such as RHEL which is fundamental in keeping your workloads more secure.

Red Hat UBI is based on RHEL, available at no additional cost, and has a minimal footprint (something between 10 **megabytes (MB)** and 100 MB, depending on the image type). There are four image types you can choose from, depending on your needs, as outlined here:

UBI Type	Usage	Highlights
`ubi-micro`	For application binaries that contain all dependencies. Applications written in Golang usually are a good fit.	Really small (~10 MB) No package manager
`ubi-minimal`	For applications that usually need a language runtime and some minimal dependencies	Contains a minimal package manager (`microdnf`)
`ubi`	For any application that runs on top of RHEL OS	Includes `openssl` and crypto stack, full yum/`dnf` package manager, and OS tools such as `gzip`, `vi`, and so on
`ubi-init`	For containers that run multiple services	Uses `systemd` to run services at startup

Besides the options mentioned, Red Hat also provides some UBIs with language runtimes included, such as Node.js, .NET, and Java. All the base images are available for download from the Red Hat Container Catalog.

Once you have a list of trusted sources that contain secure container images, we recommend you limit OpenShift to only running images from these trusted sources. You'll see next how to do this.

Limiting image registry sources

You can easily block or allow image registries that OpenShift can use with the following procedure:

1. Run the following command to edit the `image.config.openshift.io/cluster` object:

   ```
   $ oc edit image.config.openshift.io/cluster
   ```

2. In the object's YAML, set the `allowedRegistriesForImport` and `registrySources` fields with the desired image registries. `AllowedRegistriesForImport` defines the image registries to which a user can import images using image streams (the `oc import-image` command). In the `registrySources` block, you can define allowed registries (`allowedRegistries`) or blocked registries (`blockedRegistries`), but not both. Have a look at the following example:

   ```
   spec:
     allowedRegistriesForImport:
       - domainName: quay.io
         insecure: false
   ```

```
additionalTrustedCA:
  name: myconfigmap
registrySources:
  allowedRegistries:
  - quay.io
  - registry.redhat.io
  - image-registry.openshift-image-registry.svc:5000
  insecureRegistries:
  - insecure-reg.com
```

This concludes the main security aspects that we believe any company needs to think about. We encourage you to carefully review the topics we covered in this chapter and define/implement policies to keep all your clusters secure.

Summary

We have seen in this chapter some of the things you may consider to have secure OpenShift clusters. While OpenShift is considered a secure platform based on Kubernetes, how you manage it, of course, matters a lot. Consider establishing policies for access control, certificates, container and network isolation, development, and CI/CD pipelines.

A secure platform starts with proper planning to define and implement efficient policies to control what is being developed and deployed into the platform, protect it against unnecessary risks, and—finally—isolate workloads and networks as much as possible to mitigate the impacts that a breach could have on the environment. In *Chapter 11, OpenShift Multi-Cluster GitOps and Management*, you will see how to use Advanced Cluster Management to apply policies at scale to several clusters and make sure that all of them are compliant and safe, no matter where they are running.

In the next chapter, we will explore how to work with Kubernetes native pipelines on OpenShift using the Tekton project, which you can use to not only have a cloud-native CI/CD process to build and deploy your application but also add an extra layer of security in your application supply chain.

Further reading

If you want more information related to the concepts we covered in this chapter, check out the following references:

- *Is the Cloud Secure?—Gartner* study: https://www.gartner.com/smarterwith-gartner/is-the-cloud-secure

- *Kubernetes vulnerabilities and attacks reported*:

  - `https://thenewstack.io/unfixable-kubernetes-security-hole-means-potential-man-in-the-middle-attacks/`

  - `https://threatpost.com/kubernetes-cyberattacks-argo-workflows/167997/`

- *Most Common Kubernetes Security Issues and Concerns to Address* (blog post): `https://cloud.redhat.com/blog/most-common-kubernetes-security-issues-and-concerns-to-address`

- *Container Security* report: `https://www.kuppingercole.com/reprints/b1e948f62d5394353f996e43a89cde4a#heading8.1`

- *State of Kubernetes security report*: `https://www.redhat.com/en/resources/kubernetes-adoption-security-market-trends-overview`

- *OpenShift authentication guide*: `https://docs.openshift.com/container-platform/latest/authentication/index.html`

- *Troubleshooting OpenShift authentication*: `https://access.redhat.com/articles/5900841`

- *Defining default network policies*: `https://docs.openshift.com/container-platform/4.10/networking/network_policy/default-network-policy.html`

- *Red Hat Universal Base Images*: `https://developers.redhat.com/products/rhel/ubi`

- *Container Health Index grades*: `https://access.redhat.com/articles/2803031`

- *Detailed information about* htpasswd (Apache project): `https://httpd.apache.org/docs/2.4/programs/htpasswd.html`

- *OpenID Connect*: `https://developers.google.com/identity/protocols/oauth2/openid-connect`

- *Container isolation*: `https://www.nginx.com/blog/what-are-namespaces-cgroups-how-do-they-work/`

Part 3 –
Multi-Cluster CI/CD on
OpenShift Using GitOps

In this part, you will be introduced to new concepts and technologies related to CI/CD and GitOps. You will learn how to build CI/CD pipelines that are resilient to your enterprise strategy and designed to easily be maintained using Tekton and ArgoCD. You will also see how to deploy an application into multiple clusters at the same time as using Argo CD and Advanced Cluster Management.

This part of the book comprises the following chapters:

- *Chapter 9, OpenShift Pipelines – Tekton*
- *Chapter 10, OpenShift GitOps – Argo CD*
- *Chapter 11, OpenShift Multi-Cluster GitOps and Management*

9

OpenShift Pipelines – Tekton

So far in this book, we've already discussed the challenges related to the current hybrid cloud world and covered aspects regarding the OpenShift architecture and deployment. Now, we are going to shift gears and bring you an exciting DevOps-related feature: **OpenShift Pipelines**!

OpenShift Pipelines is a Kubernetes-native **continuous integration and continuous delivery (CI/CD)** tool based on the Tekton open source project, which is included at *no additional cost with Red Hat's OpenShift subscription*. In this chapter, we will walk you through it and learn how to install and use it. By doing this, you will understand how it can be helpful in your DevOps pipelines and automation.

After *Chapter 5*, *OpenShift Deployment*, you should have an OpenShift cluster working in your environment. We will use that cluster in this chapter to implement some exercises. If you don't have an OpenShift cluster available, then you can use **CodeReady Containers** (**CRC**) as a lab. Here, you can run an OpenShift cluster locally and quickly using an all-in-one VM.

In this chapter, we will cover the following topics:

- What is OpenShift Pipelines?
- Installing OpenShift Pipelines
- Creating a Tekton pipeline from scratch
- Using triggers with GitHub webhooks
- Fixing the failed PipelineRun due to YAML issues

Technical requirements

As we mentioned previously, OpenShift Pipelines is a Kubernetes native application and, as such, is a lightweight tool that uses **Custom Resource Definitions** (**CRDs**) to extend the OpenShift API's functionalities. In the upcoming sections, you will see that the installation is fairly simple and only involves installing an operator – a *"Next, Next, Finish"* sort of experience. To be able to install it and run the exercises in this chapter, you only need an OpenShift cluster with the following available resources:

- 2 vCPUs
- 2 GB of RAM

If you don't have an OpenShift cluster available to use, we recommend that you try CRC to spin up a cluster locally on your machine. To use CRC, you need to have the following system requirements on your workstation:

- 4 physical CPU cores (AMD64 or Intel 64)
- 9 GB of free memory
- 35 GB of storage space
- One of the following operating systems:

  - Windows (Windows 10 Fall Creators Update or later)
  - macOS (10.14 Mojave or later)
  - Linux (Red Hat Enterprise Linux/CentOS 7.5 or later and on the latest two stable Fedora releases)
  - Linux (Ubuntu 18.04 LTS or newer and Debian 10 or newer *are not officially* supported and may require you to manually set up the host machine

The source code used in this chapter is available at `https://github.com/PacktPublishing/OpenShift-Multi-Cluster-Management-Handbook/tree/main/chapter09`.

In this section, we will demonstrate how to install and use CRC using a Linux (Fedora) workstation. Please refer to the following site to find out more about the installation process on Windows or macOS: `https://crc.dev/crc/`.

> **What Is a CRD?**
> A CRD is a Kubernetes resource that allows you to expand the Kubernetes APIs by defining custom entities. A CRD is composed of a name and a schema that specify the API's properties.

Installing and using CRC

The CRC installation process is simple – you need to have the following packages installed in your box:

```
$ sudo yum install NetworkManager libvirt -y
```

To install CRC, follow these steps:

1. Download the latest release of CRC for your platform at `https://console.redhat.com/openshift/create/local`.

2. Extract the contents of the archive.

3. In a terminal, go to the path where you extracted the archive.

4. Run the following command to set up CRC:

    ```
    $ ./crc setup
    ```

5. If you want to set up parameters, such as the amount of CPU and memory that's available for CRC, run the following code:

    ```
    $ ./crc config set cpus 4
    $ ./crc config set memory 20480
    ```

6. Start CRC by running the following command:

    ```
    $ ./crc start
    ```

    It is going to take up to 20 minutes to completely start the cluster. At the end of the process, you will see a screen similar to the following:

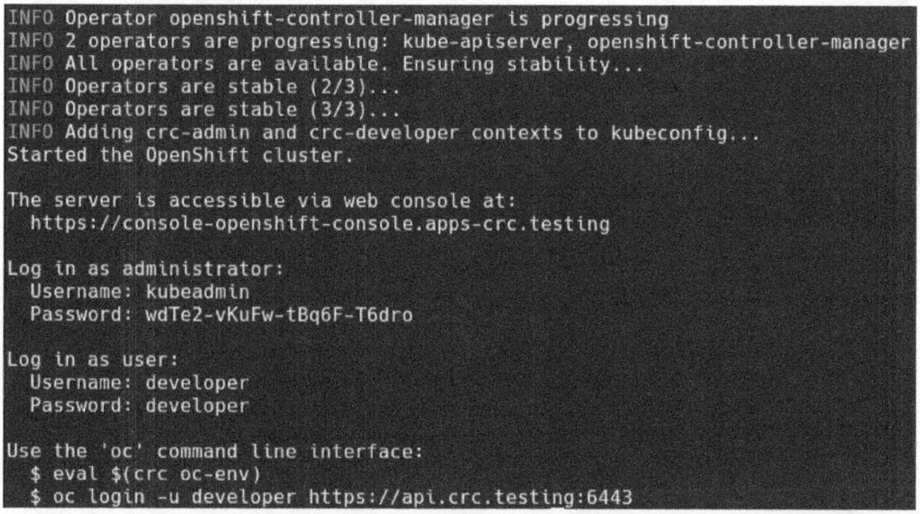

```
INFO Operator openshift-controller-manager is progressing
INFO 2 operators are progressing: kube-apiserver, openshift-controller-manager
INFO All operators are available. Ensuring stability...
INFO Operators are stable (2/3)...
INFO Operators are stable (3/3)...
INFO Adding crc-admin and crc-developer contexts to kubeconfig...
Started the OpenShift cluster.

The server is accessible via web console at:
  https://console-openshift-console.apps-crc.testing

Log in as administrator:
  Username: kubeadmin
  Password: wdTe2-vKuFw-tBq6F-T6dro

Log in as user:
  Username: developer
  Password: developer

Use the 'oc' command line interface:
  $ eval $(crc oc-env)
  $ oc login -u developer https://api.crc.testing:6443
```

Figure 9.1 – CRC startup

Now that you have CRC or any other OpenShift cluster up and running, we are ready to introduce OpenShift Pipelines and learn what you can do with it.

What is OpenShift Pipelines?

Now that you already have a lab environment, let's start our engines and drive through OpenShift Pipelines! As we mentioned previously, OpenShift Pipelines is Red Hat's implementation of the Tekton open source project. Let's learn what Tekton is and how it differs from other CI/CD pipeline tools on the market.

What is Tekton?

Tekton provides a framework for creating Kubernetes native CI/CD pipelines quickly and easily. It uses CRDs to extend the Kubernetes APIs functionalities and add some custom objects that are used to implement CI/CD pipelines. You can also integrate Tekton with industry-standard CI/CD pipeline tools such as Jenkins, GitLab CI, and any others to use the best technology for each case.

Tekton is a part of the **Continuous Delivery Foundation**, which is sponsored by huge companies such as AWS, Red Hat, Google, Netflix, and many others. This is usually a good indication that a project will have a long life and stability – an important factor for an enterprise's investment decisions.

Main benefits

Using Tekton can bring you many benefits, such as the following:

- Tekton can be considered as a serverless CI/CD pipeline system that consumes resources on demand using isolated containers, which likely reduce the infrastructure or cloud costs associated with the CI/CD tool.

- It is tightly integrated with Kubernetes, working as an extension of it using CRDs. This means that you don't need to spend time and resources with complex integrations between the CI/CD tools and OpenShift.

- Both the aforementioned aspects also mean that you will not need additional human resources to deploy, support, and maintain the CI/CD tool.

- As a Kubernetes native tool, you can define and run pipelines by applying a simple YAML file to Kubernetes (the same way you would do to create a Pod, Service, Or Deployment). This makes Tekton easy to use and integrate with other tools for complex pipelines that are composed of several components (legacy VMs, containers, microservices, and so on).

- By integrating Tekton with **Argo CD**, you can have a really powerful stack in which Tekton resides on the *continuous integration* side while Argo CD is responsible for the *continuous delivery* side. We will look at Argo CD in detail in *Chapter 10, OpenShift GitOps – Argo CD*.

- It is a true open source solution that's backed by a strong foundation, which is good evidence that it will be supported and evolve for years to come.

Tekton components

In this section, we will walk through each of the Tekton components. In a nutshell, the main Tekton components are as follows:

- **Tekton Pipelines**: It is composed of several CRDs, which are the building blocks for developing and running CI/CD pipelines.

- **Tekton Triggers**: These are objects that listen to events and trigger a pipeline or task. They are often used to run a pipeline after a pull or push request in a GitHub repository.

- **Tekton CLI**: The **command-line interface (CLI)** (tkn) to interact with Tekton.

- **Tekton Catalog**: A community-driven repository of tasks ready to be used in your pipelines.

- **Tekton Operator**: This is used to easily install, manage, and remove Tekton from a Kubernetes cluster.

Concepts

To learn Tekton, you need to understand some concepts first:

- **Step**: An action that has a set of inputs and produces a set of outputs.

- **Task**: A set of structured steps required to run a specific task, such as cloning a GitHub repository or building source code.

- **Pipeline**: A set of structured tasks that composes a CI/CD pipeline.

- **TaskRun**: This object represents the instantiation of a task. While the task is the generic definition of it, a TaskRun defines the input parameters and the other components that are needed to run it.

- **PipelineRun**: This is similar to a TaskRun, but for pipelines.

To dive into these concepts, we will cover an example where we will build and run a meaningful pipeline.

Installing OpenShift Pipelines

The installation process is really simple, as you will see in the following steps.

Prerequisites

1. You must have access to the OpenShift cluster with cluster-admin permissions.

Installation

Follow these steps:

1. Access the **OpenShift Web Console** from the administrator's perspective.

2. Navigate to **Operators | OperatorHub**:

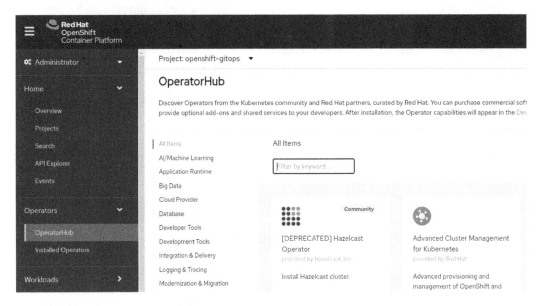

Figure 9.2 – OperatorHub

3. Search for OpenShift Pipelines using the *Filter by keyword* box:

OperatorHub

Discover Operators from the Kubernetes community and Red Hat partners, curated by Red Hat. You can purchase provide optional add-ons and shared services to your developers. After installation, the Operator capabilities will

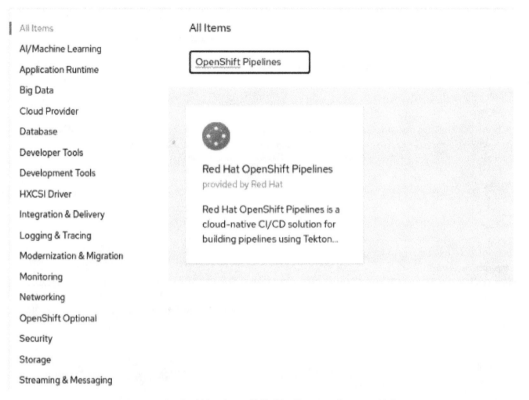

All Items	All Items
AI/Machine Learning	
Application Runtime	OpenShift Pipelines
Big Data	
Cloud Provider	
Database	
Developer Tools	
Development Tools	Red Hat OpenShift Pipelines
HXCSI Driver	provided by Red Hat
Integration & Delivery	Red Hat OpenShift Pipelines is a
Logging & Tracing	cloud-native CI/CD solution for
Modernization & Migration	building pipelines using Tekton...
Monitoring	
Networking	
OpenShift Optional	
Security	
Storage	
Streaming & Messaging	

Figure 9.3 – Red Hat OpenShift Pipelines on OperatorHub

4. Click on the **Red Hat OpenShift Pipelines** tile and then the **Install** button to see the **Install Operator** screen:

Red Hat OpenShift Pipelines
1.5.2 provided by Red Hat

Install

Latest version

1.5.2

Capability level

✅ Basic Install

✅ Seamless Upgrades

○ Full Lifecycle

○ Deep Insights

○ Auto Pilot

Source

Red Hat

Provider

Red Hat

Infrastructure features

Disconnected

Proxy-aware

Repository

https://github.com/ope
nshift/tektoncd-
operator ☑

Container image

registry.redhat.io/openshif

Red Hat OpenShift Pipelines is a cloud-native continuous integration and delivery (CI/CD) solution for building pipelines using Tekton. Tekton is a flexible Kubernetes-native open-source CI/CD framework, which enables automating deployments across multiple platforms (Kubernetes, serverless, VMs, etc) by abstracting away the underlying details.

Features

- Standard CI/CD pipelines definition
- Build images with Kubernetes tools such as S2I, Buildah, Buildpacks, Kaniko, etc
- Deploy applications to multiple platforms such as Kubernetes, serverless and VMs
- Easy to extend and integrate with existing tools
- Scale pipelines on-demand
- Portable across any Kubernetes platform
- Designed for microservices and decentralized team
- Integrated with OpenShift Developer Console

Installation

Red Hat OpenShift Pipelines Operator gets installed into a single namespace (openshift-operators) which would then install *Red Hat OpenShift Pipelines* into the openshift-pipelines namespace. *Red Hat OpenShift Pipelines* is however cluster-wide and can run pipelines created in any namespace.

Components

- Tekton Pipelines: v0.24.3
- Tekton Triggers: v0.14.2
- ClusterTasks based on Tekton Catalog 0.24

Getting Started

Figure 9.4 – Installing OpenShift Pipelines

5. Now, select **All namespaces on the cluster (default)** for **Installation mode**. As such, the operator will be installed in the openshift-operators namespace and permits the operator to install OpenShift Pipelines instances in any target namespace.

6. Select **Automatic** or **Manual** for the upgrade's **Approval Strategy**. If you go for **Automatic**, upgrades will be performed automatically by the **Operator Lifecycle Manager (OLM)** as soon as they are released by Red Hat, while if you go for **Manual**, you need to approve it before it's applied.

7. Select an **Update channel option**. The stable channel is recommended as it contains the latest stable and *supported* version of the operator.

8. Click the **Install** button:

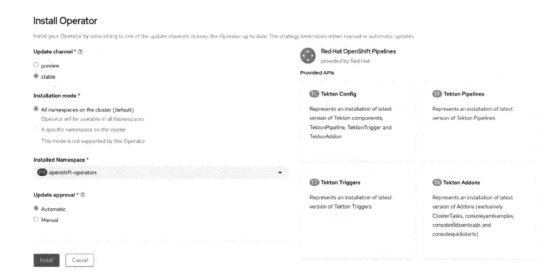

Figure 9.5 – Installing the operator

9. Wait up to 5 minutes until you see the following message:

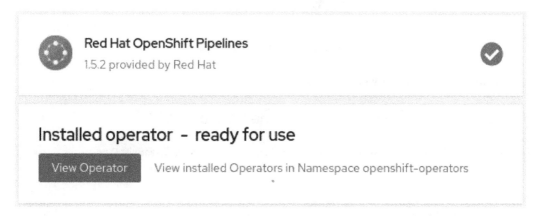

Figure 9.6 – Operator installed

Once you have installed OpenShift Pipelines, we recommend that you install the tkn CLI to help with ordinary tasks. Let's learn how to install the tkn CLI.

Installing the tkn CLI

tkn is a CLI that makes it easier to work with Tekton. Through it, you can manage (list, delete, describe, get logs, and so on) tasks, pipelines, triggers, and all the available Tekton objects.

To install the tkn CLI, follow these steps:

1. Download tkn from the URL link provided after you click the *question mark* icon of your OpenShift Web Console, as shown here:

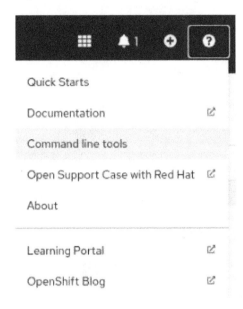

Figure 9.7 – Help menu | Command line tools

2. Download the client for your workstation:

tkn – OpenShift Pipeline Command Line Interface (CLI)

The OpenShift Pipeline client tkn is a CLI tool that allows you to manage OpenShift Pipeline resources.

- Download tkn for Linux
- Download tkn for Mac
- Download tkn for Windows
- Download tkn for IBM Power
- Download tkn for IBM Z

Figure 9.8 – tkn download links

3. After downloading it to your machine, you need to decompress it and add it to your path:

```
$ tar -xvzf tkn-linux-amd64-0.17.2.tar.gz
$ sudo cp tkn /usr/local/bin
```

4. If everything went well, you will see the output below by running `tkn version`. Ignore the warning message you will see that specifies the pipeline version; it is an expected message as we haven't logged any OpenShift clusters yet:

```
$ tkn version
Client version: 0.17.2
Pipeline version: unknown, pipeline controller may be
installed in another namespace please use tkn version -n
{namespace}
```

Now that you have installed OpenShift Pipelines and `tkn`, let's use them to create a pipeline from scratch. In the next section, we will learn about Tekton's main concepts while taking a practical approach.

Creating a Tekton pipeline from scratch

In this section, we will create a Tekton pipeline from scratch so that we can learn from it. We are going to use a sample from our GitHub repository: `https://github.com/PacktPublishing/OpenShift-Multi-Cluster-Management-Handbook`. To practice the concepts we will cover here, fork this repository to your GitHub account and follow the instructions provided in this chapter.

The pipeline that we will work on is simple but helpful. It will consist of the following tasks:

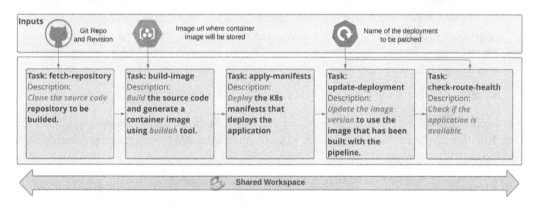

Figure 9.9 – Build and deploy pipeline

In the next few sections, you will learn how to use Tasks, TaksRuns, Pipelines, and PipelineRuns, which are Tekton's main objects.

Tasks

To create this pipeline, you need to understand the foundational concept of a task. As we mentioned previously, a task provides a set of structured steps for performing a certain action, such as cloning a GitHub repository or building source code. Now, let's go deeper and learn about some important aspects of it. The first important aspect that you need to understand is the task scope, which defines whether you need to use a Task or a ClusterTask:

- **Task**: A task is only available within a specific namespace. You will usually use tasks for actions that apply specifically to a certain application.

- **ClusterTask**: This is identical to a task but can be used in any namespace. They are usually used with generic actions that can be applied to any application.

In our example, we will use Tasks and ClusterTasks to understand how they work and the differences between them. A Task has the following elements. We will use these in our example:

- **Parameters**: The parameters that are required to run the task.

- **Resources**: This includes the input or output resources that are supplied by **PipelineResources** objects. We recommend that you use workspaces instead of PipelineResources since they are more difficult to troubleshoot, making tasks less reusable and more limited than workspaces. Due to that, we won't be using PipelineResources in our example.

- **Steps**: This is where you define the actions that will be performed in a task. You need to use a container image to run the actions.

- **Workspaces**: This is an artifact that's used to define a commonly shared storage volume between different tasks in a pipeline. Workspaces can be used for different purposes, such as sharing data between different tasks, a mount point for configurations (using ConfigMaps), credentials, and sensitive data (with secrets), and also to store reusable artifacts that have been shared between different tasks and pipelines. Workspaces are also helpful for caching artifacts to speed up builds and other jobs.

- **Results**: These are string result variables that can be passed to other tasks in a pipeline.

In our sample pipeline, we are going to reuse our existing tasks for the GitHub clone and build the source code. The last two tasks we will look at will be custom ones that we will create specifically for our pipeline.

Reusing tasks

First, let's learn how to search for and reuse tasks to build a pipeline.

The first place where you can look for existing tasks is your local OpenShift cluster. When you install OpenShift Pipelines, several **ClusterTasks** are installed with it. To check those ClusterTasks, you can use the tkn CLI or the OpenShift UI.

The following code shows how to use `tkn`:

```
# You need to login at the cluster first using "oc login"
$ oc login -u <user> https://<ocp-api-url>:6443
$ tkn clustertasks ls
```

The following is some example output:

```
$ oc login -u kubeadmin https://api.crc.testing:6443
(.. omitted ..)
$ tkn clustertasks ls
NAME                DESCRIPTION              AGE
buildah             Buildah task builds...   2 days ago
buildah-1-5-0       Buildah task builds...   2 days ago
git-cli             This task can be us...   2 days ago
git-clone           These Tasks are Git...   2 days ago
(.. omitted ..)
```

To do the same using the OpenShift UI, go to **Pipelines** | **Tasks** using the Administrator Web Console and click on the **ClusterTasks** tab. You will see the same list that we found previously using the `tkn` CLI:

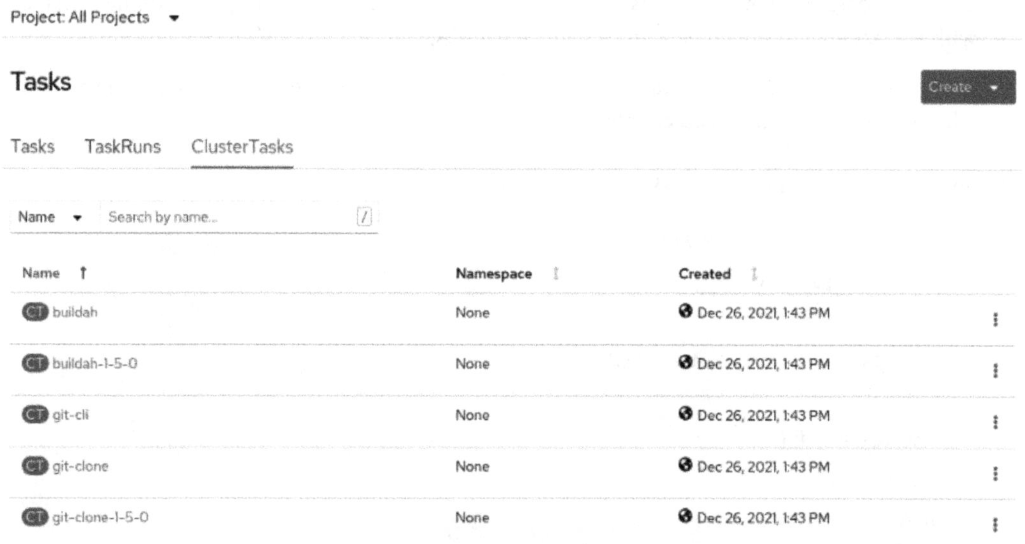

Project: All Projects ▼

Tasks

Create ▼

Tasks   TaskRuns   ClusterTasks

Name ▼   Search by name...   [/]

Name ↑	Namespace	Created
CT buildah	None	❸ Dec 26, 2021, 1:43 PM
CT buildah-1-5-0	None	❸ Dec 26, 2021, 1:43 PM
CT git-cli	None	❸ Dec 26, 2021, 1:43 PM
CT git-clone	None	❸ Dec 26, 2021, 1:43 PM
CT git-clone-1-5-0	None	❸ Dec 26, 2021, 1:43 PM

Figure 9.10 – ClusterTasks for reuse

Another great tool to look for and reuse existing tasks is **Tekton Hub**. We will use it shortly to extend our sample and validate our YAML files using the **YAML Lint** tool.

> **Notes**
>
> **Tekton Hub** is a web portal where you can get reusable assets from Tekton Catalog. It can be accessed at `https://hub.tekton.dev/`.
>
> **YAML Lint** is a tool that validates YAML file syntax, checking indentation, trailing spaces, and many other possible issues. Go to `https://yamllint.readthedocs.io/en/stable/` to learn more.

Using the ClusterTasks, we have decided to reuse the following:

- `git-clone`: To clone the source code from the GitHub repository
- `buildah`: To build the source code and generate a container image as a result

Now, let's learn how to create a custom task for when you need something specific.

Creating a new (custom) task

Defining a new task is as simple as creating a pod or deployment. For our example, we need to create three new tasks:

- `apply-manifests`: This task will be responsible for applying some K8s manifest files that will deploy the application.
- `update-deployment`: This task will update the deployment, replacing the container image with the one that has been built in the previous tasks.
- `check-app-health`: This task checks the application pod's status and the URL to validate whether the application is accessible.

Let's create these tasks, explore their content, and learn from them:

```
apiVersion: tekton.dev/v1beta1
kind: Task #[1]
metadata:
  name: apply-manifests #[2]
spec:
  workspaces: #[3]
  - name: source
  params: #[4]
  - name: manifest_dir
    description: The directory in the source that contains yaml
manifests
    type: string
```

```
        default: "k8s"
    steps: #[5]
       - name: apply
         image: image-registry.openshift-image-registry.svc:5000/
openshift/cli:latest #[6]
         workingDir: /workspace/source
#[7]
         command: ["/bin/bash", "-c"]
         args:
           - |-
           echo Applying manifests in $(inputs.params.manifest_
dir) directory
             oc apply -f $(inputs.params.manifest_dir)
             echo ----------------------------------
```

In the preceding code, we have highlighted some parts with numbers. Let's take a look:

- **[1]**: An object of this kind defines a new Tekton task.

- **[2]**: The task's name.

- **[3]**: The workspace containing the k8s manifest source files. This is the shared workspace that was populated with the git-clone task (the first task).

- **[4]**: The parameters that are required for the task to run.

- **[5]**: The steps that are performed with the task.

- **[6]**: The image that will run the step commands.

- **[7]**: The commands that will perform the desired action – in this case, applying the k8s manifests.

Now that we have looked at the task's structure, let's create it in our sample environment and run it using another object – **TestRun**:

1. Create a new project for our example:

```
$ oc new-project pipelines-sample
```

2. Now, check if the pipeline's service account has been created automatically:

```
$ oc get serviceaccount pipeline
NAME        SECRETS    AGE
pipeline    2          33s
```

3. Create the `apply-manifest` task in the `pipelines-sample` namespace:

```
$ oc apply -f  https://github.com/PacktPublishing/
OpenShift-Multi-Cluster-Management-Handbook/blob/main/
chapter09/Tasks/apply-manifests.yaml
task.tekton.dev/apply-manifests created
```

4. Using `tkn` confirm that the task has been created:

```
$ tkn tasks ls
NAME                DESCRIPTION    AGE
apply-manifests                    17 seconds ago
```

5. Now, let's create other custom tasks (`update-image-version` and `check-route-health`):

```
$ oc apply -f  https://github.com/PacktPublishing/
OpenShift-Multi-Cluster-Management-Handbook/blob/main/
chapter09/Tasks/update-image-version.yaml
$ oc apply -f  https://github.com/PacktPublishing/
OpenShift-Multi-Cluster-Management-Handbook/blob/main/
chapter09/Tasks/check-route-health.yaml
$ tkn tasks ls
NAME                DESCRIPTION    AGE
apply-manifests                    17 seconds ago
heck-app-health                    10 seconds ago
update-deployment                  8 seconds ago
```

Now that we have created our custom tasks, let's learn how to run and test them using a `TaskRun` object.

TaskRun

Our task needs a persistent volume to store the source code from GitHub. As such, before we run TaskRun, we need to create a **PersistentVolumeClaim**. Note that you need to have a **StorageClass** to provision a **PersistentVolume** automatically for you. If you don't have one, the PersistentVolumeClaim will be *Pending*, waiting for the PersistentVolume to be created manually.

Run the following command to create the PersistentVolumeClaim:

```
$ oc apply -f https://github.com/PacktPublishing/OpenShift-
Multi-Cluster-Management-Handbook/blob/main/chapter09/
PipelineRun/pvc.yaml
```

Now, we must create two TaskRuns. In the first, we will use the `git-clone` ClusterTask to clone the GitHub repository and store it in the workspace that uses a PersistentVolume. In the second, we will use the custom task that we created previously, which deploys the application by applying some manifests (the `apply-manifests` task).

The following code shows the structure of a TaskRun:

```
apiVersion: tekton.dev/v1beta1
kind: TaskRun
metadata:
  name: git-clone #[1]
spec:
  taskRef:
    name: git-clone #[2]
    kind: ClusterTask #[3]
  params: #[4]
  - name: url
    value: "https://github.com/PacktPublishing/OpenShift-Multi-
Cluster-Management-Handbook"
  - name: subdirectory
    value: ""
  - name: deleteExisting
    value: "true"
  - name: revision
    value: "main"
  workspaces: #[5]
  - name: output
    persistentVolumeClaim:
      claimName: source-pvc
```

Let's look at this code in more detail:

- **[1]**: The name of the `TaskRun` object
- **[2]**: The name of the task that will be run
- **[3]**: This is required for `ClusterTask` but can be omitted for regular tasks
- **[4]**: Parameter values to be used during the task's execution
- **[5]**: Workspaces to be used

Run the following command to apply the `TaskRun` object:

```
$ oc apply -f https://github.com/PacktPublishing/OpenShift-
Multi-Cluster-Management-Handbook/blob/main/chapter09/Tasks/
git-clone-taskrun.yaml
```

Once you have created the `git-clone` object, you can look at the logs using the following `tkn` command:

```
$ tkn taskrun logs git-clone -f
[clone] + '[' false = true ']'
[clone] + '[' false = true ']'
[clone] + CHECKOUT_DIR=/workspace/output/
[clone] + '[' true = true ']'
[clone] + cleandir
[clone] + '[' -d /workspace/output/ ']'
(.. ommited ..)
```

Finally, run `apply-manifests` using the following TaskRun:

```
$ oc apply -f https://github.com/PacktPublishing/OpenShift-
Multi-Cluster-Management-Handbook/blob/main/chapter09/Tasks/
apply-manifests-taskrun.yaml
```

Check the logs, as follows:

```
$ tkn taskrun logs run-apply-manifests -f
[apply] Applying manifests in ./sample-go-app/articles-api/k8s
directory
[apply] deployment.apps/clouds-api created
[apply] service/clouds-api created
[apply] route/clouds-api created
```

With that, you have learned how to run a particular task using a `TaskRun` object. You also know how to reuse and create a custom task. We will use this knowledge to create our first pipeline.

Pipelines

In this section, we will create our first meaningful pipeline! I like to compare a pipeline's design to a LEGO® set, in which you need to have all the pieces at hand before assembling it. If the LEGO set is too big to assemble at once, you need to break it into smaller blocks of meaningful parts. In our pipeline, the *LEGO pieces* are the **tasks** that we have already built and the ones we will reuse. We have all we need, so *let's assemble our LEGO set*.

We will use our example to understand how to define a pipeline object. The first part of any pipeline is its metadata:

```
apiVersion: tekton.dev/v1beta1
kind: Pipeline
metadata:
  name: build-and-deploy
```

The next part is its specification (spec), which is composed of the following items:

- **Workspaces**: This is a shared workspace that's required to store the source code and any other pipeline artifacts that need to be passed between the tasks:

```
spec:
  workspaces:
  - name: shared-workspace
```

- **Parameters**: These are the input parameters that are required to run the pipeline:

```
params:
- name: deployment-name
  type: string
  description: name of the deployment to be patched
- name: git-url
  type: string
  description: url of the git repo for the code of
deployment
  - name: git-revision
  type: string
  description: revision to be used from repo of the
code for deployment
  default: "master"
- name: IMAGE
  type: string
  description: image to be built from the code
```

- **Tasks**: These are the tasks to be run. Each task must have a valid name and a taskRef (the reference to the task that will be used), as follows:

```
- name: apply-manifests
  taskRef:
```

```
      name: apply-manifests
  workspaces:
  - name: source
    workspace: shared-workspace
  runAfter:
  - build-image
```

- For **ClusterTasks**, you need to explicitly set the `kind` attribute within the `taskRef` group, like so:

```
- name: fetch-repository
  taskRef:
    name: git-clone
    kind: ClusterTask
  workspaces:
  - name: output
    workspace: shared-workspace
  params:
  - name: url
    value: $(params.git-url)
  - name: subdirectory
    value: ""
  - name: deleteExisting
    value: "true"
  - name: revision
    value: $(params.git-revision)
```

You can find the complete pipeline at `https://github.com/PacktPublishing/OpenShift-Multi-Cluster-Management-Handbook/blob/main/chapter06`.

Now, we are ready to create our pipeline. To do so, run the following command:

```
$ oc apply -f https://github.com/PacktPublishing/OpenShift-
Multi-Cluster-Management-Handbook/blob/main/chapter09/Pipeline/
build-deploy.yaml
$ tkn pipelines ls
NAME                AGE            LAST
RUN    STARTED   DURATION    STATUS
build-and-deploy    1 minute ago   ---          ---
```

Now that we have defined our pipeline, let's run it!

PipelineRun

There are multiple ways to run the pipeline: through the OpenShift Console UI, using `tkn`, or by creating and applying a PipelineRun object manually. At the end of the day, no matter how you run it, a PipelineRun will always be created (the only difference is that the PipelineRun is created automatically for you when you use `tkn` or the web UI). For didactic reasons, we will do this using a `PipelineRun` object to learn about and understand it.

The following code shows our `PipelineRun` object:

```
apiVersion: tekton.dev/v1beta1
kind: PipelineRun
metadata:
  name: build-deploy-api-pipelinerun #[1]
spec:
  pipelineRef:
    name: build-and-deploy #[2]
  params: #[3]
  - name: deployment-name
    value: clouds-api
  - name: git-url
    value: https://github.com/PacktPublishing/OpenShift-Multi-
Cluster-Management-Handbook.git
  - name: IMAGE
    value:  image-registry.openshift-image-registry.svc:5000/
pipelines-sample/clouds-api
  workspaces: #[4]
  - name: shared-workspace
    volumeClaimTemplate:
      spec:
        accessModes:
          - ReadWriteOnce
        resources:
          requests:
            storage: 500Mi
```

Let's look at this code in more detail:

- **[1]**: The name of the `PipelineRun` object
- **[2]**: The pipeline to be run
- **[3]**: The parameter values to be used with the pipeline
- **[4]**: The workspace's definition

Apply the `PipelineRun` object and check the logs to see the pipeline's execution:

```
$ oc apply -f https://github.com/PacktPublishing/OpenShift-
Multi-Cluster-Management-Handbook/blob/main/chapter09/
PipelineRun/clouds-api-build-deploy.yaml
$ tkn pipelinerun logs build-deploy-api-pipelinerun -f
[fetch-repository : clone] + '[' false = true ']'
[fetch-repository : clone] + '[' false = true ']'
[fetch-repository : clone] + CHECKOUT_DIR=/workspace/output/
(.. omitted ..)
[check-app-health : apply] Waiting for application articles-api
to be ready.
[check-app-health : apply] Checking if application is available
at the route endpoint
[check-app-health : apply] Application is available at http://
articles-api-pipelines-sample.apps.cluster-gf.gf.sandbox1171.
opentlc.com/cloud
[check-app-health : apply] --------------------------------
```

With that, you have custom tasks and a pipeline that has been tested and is working already. Now, let's make it even better by using a trigger to run this pipeline automatically when a Git push occurs in the repository.

Using triggers with GitHub webhooks

In a CI/CD workflow, it is typical to use an event, such as a pull or push request on Git, to trigger a new pipeline run. With Tekton, you use **EventListeners** to listen for events and run one or more triggers. There are some out-of-the-box event processors, named **Interceptors**, for the following platforms:

- **GitHub**: This allows you to validate and filter GitHub webhooks.
- **GitLab**: The same as the previous point but for GitLab.
- **Bitbucket**: The same as the previous points for Bitbucket.

- **CEL**: This allows you to use **Common Expression Language** (CEL) to filter and modify payloads.

- **Webhook**: This allows you to process any webhook payload and apply any business logic to it.

In our example, we will use a GitHub interceptor to process a webhook, filter push events, and trigger the pipeline we created previously. You can also implement your custom interceptors by implementing an object named `ClusterInterceptors`. Check out the links in the *Further reading* section if you need to create a ClusterInterceptor or use any interceptor other than the GitHub one that we will use in our example.

Note that the GitHub webhook requires a publicly accessible URL to send the HTTP webhook posts. Due to that, you will need an OpenShift cluster with a public IP and domain that can be accessed from the internet. That said, in this case, you will not be able to use CRC to test Tekton triggers using GitHub webhooks unless you make your CRC URL routes public on the internet.

> **What is CEL?**
>
> CEL is a simple but fast and portable language for expression evaluation. Created and maintained by some Google engineers, it is an open source project that was released under the Apache License and used with many Google projects and services. For more information, go to `https://opensource.google/projects/cel`.

Besides the **EventListener**, a Tekton trigger is composed of several other objects:

- **Trigger**: This defines which action will be performed after the EventListener detects a new event.

- **TriggerTemplate**: This specifies a blueprint of objects that will be applied as a result of the trigger, usually using a PipelineRun object, which, in turn, will run a pipeline.

- **TriggerBinding**: This defines the field data that will be extracted from the event payload to be used with the associated PipelineRun.

- **ClusterTriggerBinding**: This is the same as the TriggerBinding but cluster-scoped. It can be reused among different namespaces.

The following objects will be used in our example:

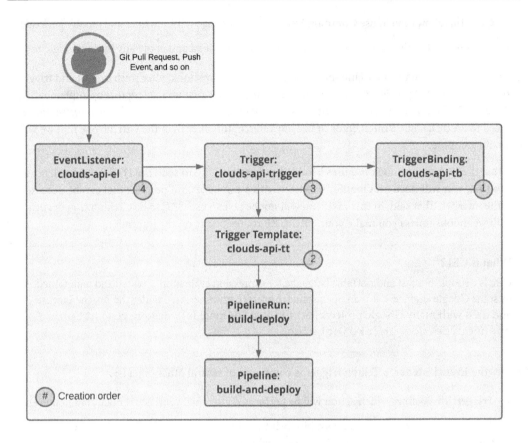

Figure 9.11 – Tekton Trigger objects

Now, let's put this into practice! You have already created the tasks and pipeline in your lab, so let's create the trigger objects that will use the existing pipeline.

TriggerBinding

TriggerBinding will parse the data that's been extracted from the GitHub payload, as follows:

```
apiVersion: triggers.tekton.dev/v1beta1
kind: TriggerBinding
metadata:
  name: clouds-api-tb #[1]
spec:
  params: #[2]
  - name: git-repo-url
```

```
      value: $(body.repository.url)
    - name: git-repo-name
      value: $(body.repository.name)
    - name: git-revision
      value: $(body.head_commit.id)
```

Let's look at this code in more detail:

- **[1]**: The name of the `TriggerBinding` object
- **[2]**: The parameters that will be assigned according to the payload data fields

Use the following command to create the TriggerBinding:

```
$ oc apply -f https://raw.githubusercontent.com/
```

The next object we need to create is `TriggerTemplate`. Let's take a look.

TriggerTemplate

`TriggerTemplate` will create a PipelineRun that executes our sample pipeline:

```
apiVersion: triggers.tekton.dev/v1beta1
kind: TriggerTemplate
metadata:
  name: clouds-api-tt #[1]
spec:
  params: #[2]
  - name: git-repo-url
    description: The git repository url
  - name: git-revision
    description: The git revision
    default: master
  - name: git-repo-name
    description: The name of the deployment to be created /
patched
  resourcetemplates: #[3]
  - apiVersion: tekton.dev/v1beta1
    kind: PipelineRun
    metadata:
      generateName: build-deploy-
```

```
    spec:
      serviceAccountName: pipeline
      pipelineRef:
        name: build-and-deploy
      params:
      - name: deployment-name
        value: clouds-api
      - name: git-url
        value: $(tt.params.git-repo-url)
      - name: git-revision
        value: $(tt.params.git-revision)
      - name: IMAGE
        value: image-registry.openshift-image-registry.
svc:5000/pipelines-sample/clouds-api
      workspaces:
      - name: shared-workspace
        volumeClaimTemplate:
          spec:
            accessModes:
              - ReadWriteOnce
            resources:
              requests:
                storage: 500Mi
```

Let's look at this code in more detail:

- **[1]**: The name of the `TriggerTemplate` object
- **[2]**: The input parameters that are populated by the `TriggerBinding` object
- **[3]**: The objects that will be created as a result of the trigger

Use the following command to create the `TriggerTemplate` object:

```
$ oc apply -f https://github.com/PacktPublishing/OpenShift-
Multi-Cluster-Management-Handbook/blob/main/chapter09/Trigger/
clouds-api-tt.yaml
```

Finally, we can create the `Trigger` object, which uses all the objects we have created.

Trigger

The `Trigger` object will be the glue between the GitHub interceptor, `TriggerBinding`, and `TriggerTemplate`:

```
apiVersion: triggers.tekton.dev/v1beta1
kind: Trigger
metadata:
  name: clouds-api-trigger #[1]
spec:
  serviceAccountName: pipeline
  interceptors: #[2]
    - ref:
        name: "github" #[3]
      params:
        - name: "secretRef" #[4]
          value:
            secretName: github-secret
            secretKey: secretToken
        - name: "eventTypes"
          value: ["push"] #[5]
  bindings:
    - ref: clouds-api-tb #[6]
  template:
    ref: clouds-api-tt #[7]
```

Let's look at this code in more detail:

- **[1]**: The name of the `Trigger` object.
- **[2]**: The list of event interceptors to be used to trigger the actions.
- **[3]**: The interceptor from GitHub.
- **[4]**: The secret that's been configured in the GitHub webhook.
- **[5]**: The trigger event types that Tekton will react to. In this case, it will be GitHub "push" events.
- **[6]**: The `TriggerBinding` object that will be used with this trigger.
- **[7]**: The `TriggerTemplate` object that will be used with this trigger.

The following code shows an example of the GitHub secret (**[4]**):

```
apiVersion: v1
kind: Secret
metadata:
  name: github-secret
type: Opaque
stringData:
  secretToken: "tekton"
```

Use the following command to create the secret and trigger:

```
$ oc apply -f https://github.com/PacktPublishing/OpenShift-
Multi-Cluster-Management-Handbook/blob/main/chapter09/Trigger/
clouds-api-trigger.yaml
```

The last object we need to create to put the trigger into practice is EventListener. Let's take a look.

EventListener

Finally, we need to create an **EventListener** object that will listen for HTTP requests and be used with the GitHub webhook configuration. We will learn how to configure the GitHub webhook soon:

```
apiVersion: triggers.tekton.dev/v1beta1
kind: EventListener
metadata:
  name: clouds-api-el #[1]
spec:
  serviceAccountName: pipeline
  triggers:
    - triggerRef: vote-trigger #[2]
```

Let's look at this code in more detail:

- **[1]**: The name of the EventListener object
- **[2]**: The trigger that will be invoked when the EventListener object is sensitized

Run the following command to create the EventListener object:

```
$ oc apply -f https://github.com/PacktPublishing/OpenShift-
Multi-Cluster-Management-Handbook/blob/main/chapter09/Trigger/
clouds-api-el.yaml
```

`EventListener` will create a service on OpenShift that you need to expose externally. The route URL that's generated will be used during the GitHub webhook configuration:

```
$ oc expose svc el-clouds-api-el
```

Now, we are ready to configure a new GitHub webhook that will use the `EventListener` object we just created to fire Tekton's trigger.

Creating a GitHub webhook

To create a webhook, you need to fork our GitHub repository. If you haven't forked it yet, do so now in your personal GitHub account: `https://github.com/PacktPublishing/OpenShift-Multi-Cluster-Management-Handbook`.

Open the GitHub forked repository and go to **Settings | Webhook**. On the following page, click on the **Add webhook** button:

Figure 9.12 – Adding a webhook on GitHub

Fill out the form by providing the following information:

- **Payload URL**: The route URL we created in the previous section. You can get this URL by running the following command:

```
$ echo "$(oc  get route el-clouds-api-el
--template='http://{{.spec.host}}')"
```

- **Content type**: `application/json`.

- **Secret**: The same value you used with the Tekton trigger's secret (in our case, this is `tekton`):

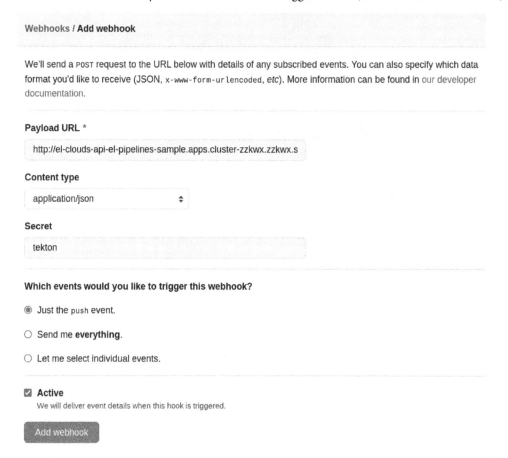

Figure 9.13 – Adding a webhook on GitHub

After a few seconds, you should see a green check mark next to the webhook we created:

Figure 9.14 – Webhook added

Now that we have our `Trigger` objects from the Tekton side and the webhook configured on GitHub, let's test it!

Testing the Tekton trigger

Run a `commit` and push the trigger to the webhook event, like so:

```
$ git commit -m "empty-commit" --allow-empty && git push origin
main
```

Access the **Pipelines** menu. You should see a new **PipelineRun** start after the Git push command has been run:

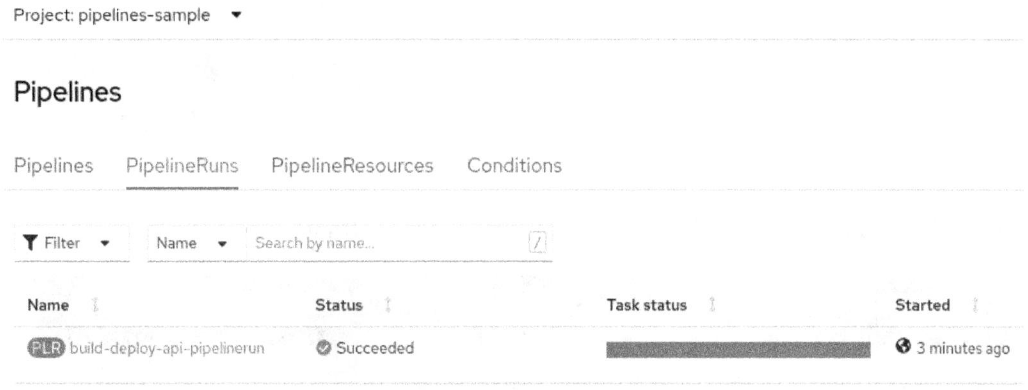

Project: pipelines-sample   ▼

Pipelines

Pipelines   PipelineRuns   PipelineResources   Conditions

▼ Filter ▼   Name ▼   Search by name...

Name	Status	Task status	Started
PLR build-deploy-api-pipelinerun	✓ Succeeded		⊘ 3 minutes ago

Figure 9.15 – PipelineRun on Red Hat

Congratulations – you have successfully created a CI/CD pipeline on Tekton and ran it automatically after a Git push event using a trigger! To wrap up this chapter, we will enhance our pipeline by adding a validation task for YAML files using the YAML linter tool.

To do so, let's use Tekton Hub to find a reusable task. Go to `https://hub.tekton.dev/` and search for YAML using the search box at the top right of the screen:

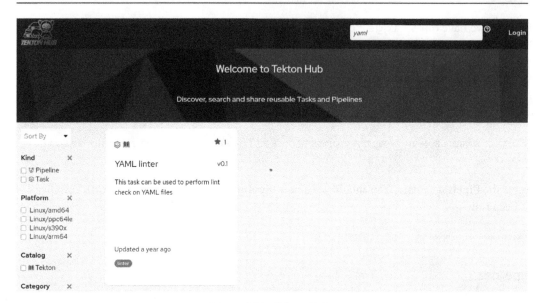

Figure 9.16 – Tekton Hub

Click the **YAML linter** task to find the instructions on how to install and use it:

Figure 9.17 – YAML linter

This time, we will use the **Pipeline Builder** page to add the YAML linter task. To do so, access the OpenShift UI and select the **Developer** console:

Figure 9.18 – Developer console

Now, perform the following steps:

1. Access the **Pipelines** menu and click on the `build-and-deploy` pipeline:

Figure 9.19 – The Pipelines menu

2. Now, click the **Actions** button and then **Edit Pipeline**:

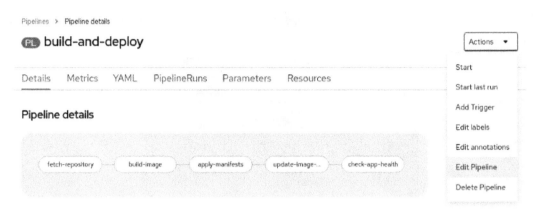

Figure 9.20 – The build-and-deploy pipeline

3.   On the following screen, click on the **fetch-repository** box and then the + sign next to it:

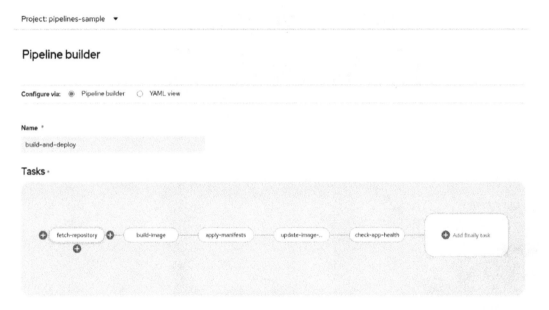

Figure 9.21 – The Pipeline builder feature

4.   Select the **Add task** box, type yaml lint, and click the **Install and add** button:

Pipeline builder

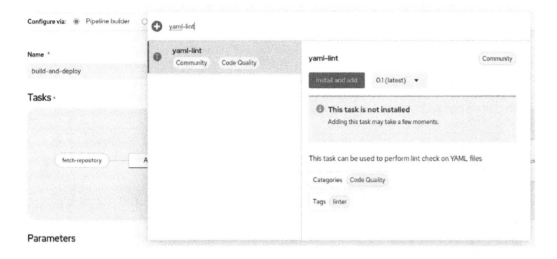

Figure 9.22 – Adding a new task using the Pipeline builder feature

5.   The new task should have been added. You should see an exclamation mark next to it:

Figure 9.23 – Adding a new task using the Pipeline builder feature

6.   Now, click it and input `./sample-go-app/clouds-api/k8s` as the **args** parameter and `shared-workspace` as the **Workspaces** group:

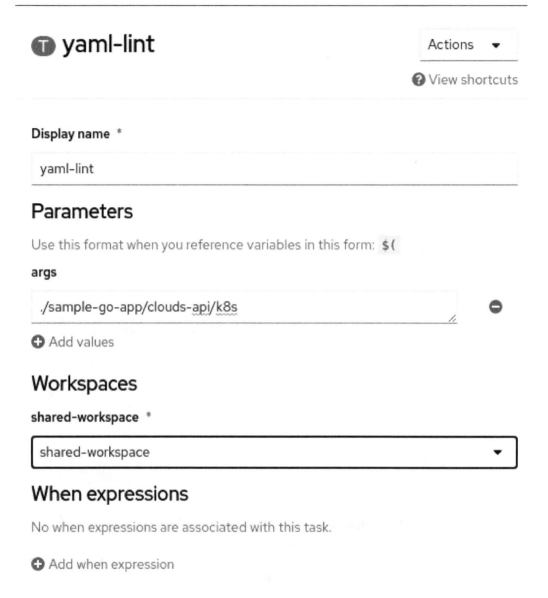

Figure 9.24 – Setting the yaml-lint task's parameters

7. Now, click on **Save**.

8. At this point, our pipeline has a new step that validates the YAML content of our Kubernetes manifest files. To test our previous change, let's run it from the same web UI. To do so, click the **Actions** menu from the **Pipeline details** screen and select the **Start** action:

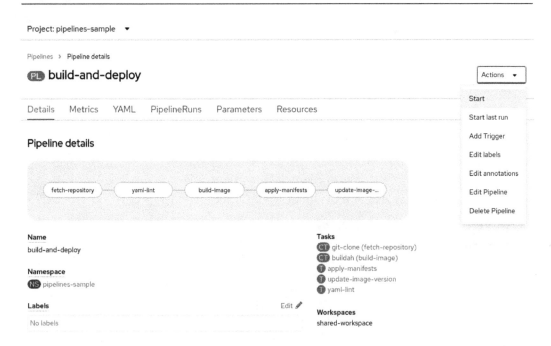

Figure 9.25 – Running the pipeline from the Developer console

9. Fill out the form by using the following values and click **Start**:

- **deployment-name**: `clouds-api`

- **git-url**: <Your forked repository>

- **git-revision**: `main`

- **IMAGE**: `image-registry.openshift-image-registry.svc:5000/pipelines-sample/clouds-api`

- **CONTEXT**: `./sample-go-app/clouds-api/`

- **shared-workspace**: `VolumeClaimTemplate:`

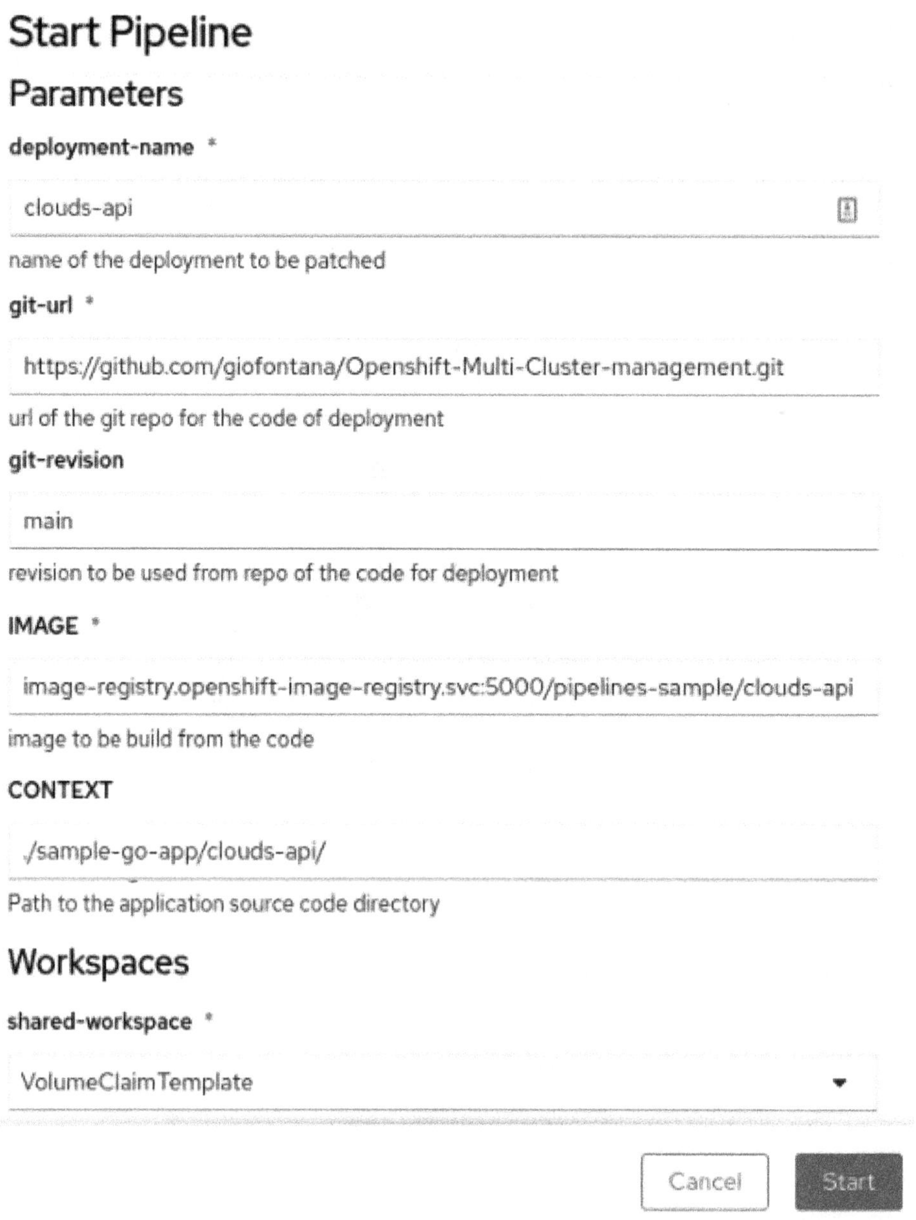

Start Pipeline

Parameters

deployment-name *

> clouds-api

name of the deployment to be patched

git-url *

> https://github.com/giofontana/Openshift-Multi-Cluster-management.git

url of the git repo for the code of deployment

git-revision

> main

revision to be used from repo of the code for deployment

IMAGE *

> image-registry.openshift-image-registry.svc:5000/pipelines-sample/clouds-api

image to be build from the code

CONTEXT

> ./sample-go-app/clouds-api/

Path to the application source code directory

Workspaces

shared-workspace *

> VolumeClaimTemplate   ▾

Cancel   Start

Figure 9.26 – PipelineRun parameters

10. Check the `PipelineRun` object on the following screen. You will get an error regarding the new **yaml-lint** task we added:

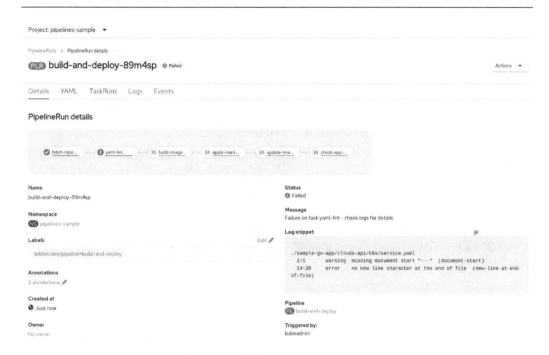

Figure 9.27 – PipelineRun failed due to YAML linter validations

11.   Click the **yaml-lint** step and check out the logs to find the issue:

Figure 9.28 – PipelineRun failed due to YAML linter validations

As you can see, the YAML linter detected errors in some YAML files. Those errors are expected and have been prepared especially for you to simulate what a real CI/CD pipeline looks like. Now, practice the skills you've just acquired to fix those errors and get the pipeline working again (or look at the solution in the next section)!

Fixing the failed PipelineRun due to YAML issues

To get your pipeline working again, follow these steps:

1. Add `---` to the first line of all the YAML files in the `./sample-go-app/clouds-api/k8s` folder.

2. Fix the indentation of the `kustomization.yaml` file by adding two spaces at the beginning of all the lines after `resources`, like so:

   ```
   ---

   resources:
     - deployment.yaml #add two spaces at the begging of the
   line
     - service.yaml #add two spaces at the begging of the
   line
     - route.yaml #add two spaces at the begging of the line
   ```

3. Add a new line at the end of the `service.yaml` file.

4. Commit and push the changes:

   ```
   $ git add *
   $ git commit -m "fixed yaml files"
   $ git push
   ```

A new PipelineRun should be triggered automatically and complete.

Summary

In this chapter, we dived into Tekton, from installing it on OpenShift to using it. You learned how to create custom tasks, reuse existing ones, build a pipeline, and then run it. You also learned how to set a trigger to run the pipeline when a push event occurs in your GitHub repository. The objects you have seen in this chapter are the main ones you will use to create most Tekton pipelines.

In the next chapter, we will bring more power to your CI/CD process by adding **Argo CD** and **GitOps** to your pipelines. We will also start looking at ways to deploy applications into multiple clusters at once. Let's get started and take a deeper dive into OpenShift GitOps!

Further reading

If you want to learn more about what we covered in this chapter, check out the following references:

- *OpenShift Pipelines official documentation:* `https://docs.openshift.com/container-platform/4.9/cicd/pipelines/understanding-openshift-pipelines.html`

- *Tekton official documentation:* `https://tekton.dev/docs/overview/`

- *How to create custom interceptors using* **ClusterInterceptor**: `https://tekton.dev/docs/triggers/clusterinterceptors/`

- *Tekton Hub (a collection of reusable tasks):* `https://hub.tekton.dev/`

OpenShift GitOps – Argo CD

In the previous chapter, we learned how to create and run a pipeline using Tekton to build and deploy an application. While Tekton is great for building and performing other actions that are usually related to **continuous integration** (**CI**), **GitOps** is becoming the norm for **continuous deployment** (**CD**) regarding Kubernetes-native applications. In this chapter, we will dive into GitOps and talk about one of the best tools for CD: **Argo CD**.

In this chapter, we will cover the following topics:

- What is GitOps?

- What is Argo CD?

- Application delivery model

- Installing OpenShift GitOps

- Configuring Argo CD against multiple clusters

- Argo CD definitions and challenges

- Argo CD main objects

- Deploying an application using GitOps

- Deploying to multiple clusters

Let's dive in!

> **Note**
> The source code used in this chapter is available at https://github.com/
> PacktPublishing/OpenShift-Multi-Cluster-Management-Handbook/
> tree/main/chapter10.

What is GitOps?

The term GitOps was first described by *Lexis Richardson*, CEO of *Weaveworks*, in 2017. At that time, he presented the four principles of GitOps, which are as follows:

- **The entire system is described declaratively**: This means that any configuration of your application and infrastructure needs to be treated as code, but not as a set of instructions, as you would with scripts or automation code. Instead, you must use a set of facts that describes the desired state of your system. These declaration files are versioned in Git, which is your single source of truth. The great benefit of this principle is that you can easily deploy or roll back your applications and, more importantly, restore your environment quickly if a disaster occurs.

- **The canonical desired system state is versioned in Git**: Git is your source of truth. It needs to be the single place that triggers all the changes in your systems. Ideally, nothing should be done directly on the systems, but through configuration changes on Git that will be applied automatically using a tool such as Argo CD.

- **Approved changes are automatically applied to the system**: Since you have the desired state of your system stored in Git, any changes can be automatically applied to the system as they are pushed to the repository.

- **Software agents ensure correctness and alert you about divergence**: It is crucial to have tools in place that will ensure that your system is in the desired state, as described in Git. If any drift is detected, the tool needs to be able to self-heal the application and get it back to its desired state.

GitOps became the norm for Kubernetes and cloud-native applications due to the following benefits:

- **Standard process and tools**: Git workflows allow teams to work collaboratively and in a reproducible manner, avoiding issues regarding human-repetitive tasks.

- **Robust and secure process**: By working with **pull requests** (**PRs**) in Git, all the changes need to be reviewed and approved. You can also trace all changes in Git and revert them if needed.

- **Auditable changes**: All changes are tracked and easily auditable in Git history.

- **Consistency**: You can deploy the same application in multiple different clusters consistently:

Declarative state

Continuous delivery

Actual state

Figure 10.1 – GitOps workflow

Now that you have a better understanding of what GitOps is, let's start learning how to put GitOps workflows into practice.

What is Argo CD?

In theory, it is possible to adopt GitOps without the need to use any specific tool. You could implement scripts and automation to deploy and manage your applications using declarative files from Git that describe your systems. However, that would be costly and time-consuming. The good news is that there are some great open source options for Kubernetes that are stable and work well. At the time of writing, the main tools for Kubernetes are **Argo CD** and **Flux CD**; both are great tools but in this book, we will explore Argo CD, which comes at *no additional cost with a Red Hat OpenShift subscription*.

In a nutshell, Argo CD is a tool that is capable of *reading* a set of *Kubernetes manifests, Helm charts, or Jsonnet files* stored in a Git repository and *applying* them to a Kubernetes namespace. Argo CD is not only able to apply manifests, though – it can also automate self-healing, object pruning, and other great capabilities, as we will explore in this chapter.

Application delivery model

At this point, you may be wondering how OpenShift Pipelines (**Tekton**) and GitOps (**Argo CD**) are related. Tekton and Argo CD are complementary tools that are perfect together. While Tekton is a perfect fit for *CI* pipelines that run unit tests and build and generate container images, Argo CD is more appropriate for *continuous delivery* practice. The following diagram summarizes what a CI/CD pipeline with Tekton and Argo CD looks like:

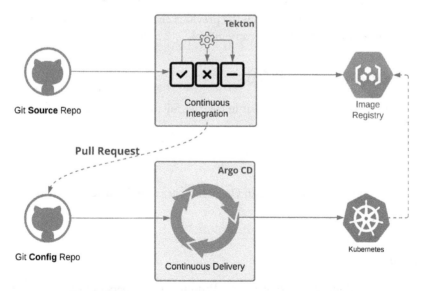

Figure 10.2 – Application delivery model using Tekton and Argo CD

CD with GitOps means that the actual state of the application should be monitored and that any changes need to be reverted to the application's desired state, as described in the Git repository:

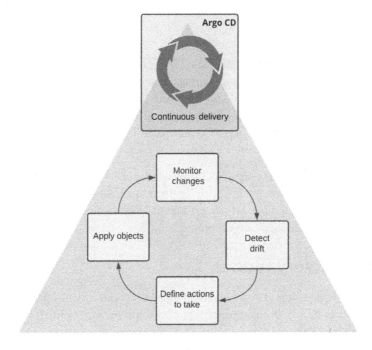

Figure 10.3 – Continuous delivery with GitOps

In this chapter, we will use our example from the previous chapter and use Argo CD to deploy the application and practice this application delivery model.

Installing OpenShift GitOps

The installation process is simple and is similar to what we followed in the previous chapter regarding OpenShift Pipelines.

Prerequisites

To install OpenShift GitOps, you will need an OpenShift cluster with cluster-admin permissions.

Installation

Follow these steps:

1. Access the **OpenShift web console** using the administrator's perspective.

2. Navigate to **Operators | OperatorHub**:

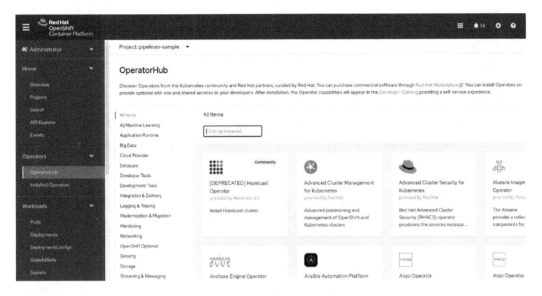

Figure 10.4 – OperatorHub

3. Search for OpenShift GitOps using the *Filter by keyword* box:

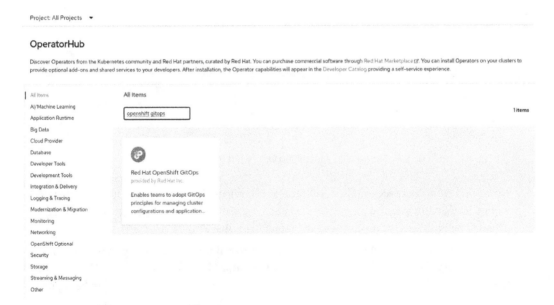

Figure 10.5 – Red Hat OpenShift GitOps on OperatorHub

4. Click on the **Red Hat OpenShift GitOps** tile and then the **Install** button to go to the **Install** screen:

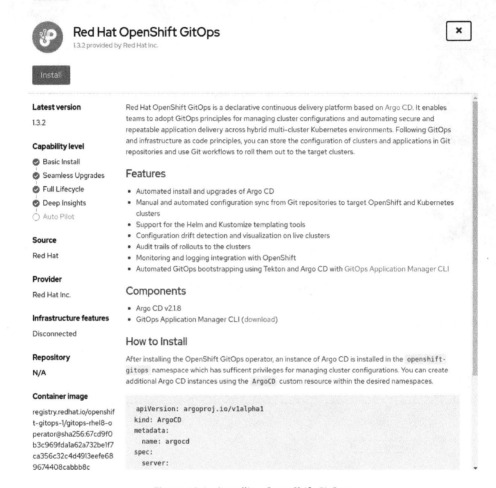

Figure 10.6 – Installing OpenShift GitOps

5. Now, select **All namespaces on the cluster (default)** for **Installation mode**. As such, the operator will be installed in the `openshift-operators` namespace and permits the operator to install OpenShift GitOps instances in any target namespace.

6. Select **Automatic** or **Manual** for the upgrade's **Approval Strategy**. If you go for **Automatic**, upgrades will be performed automatically by the **Operator Lifecycle Manager** (**OLM**) as soon as they are released by Red Hat, while for **Manual**, you need to approve it before it can be applied.

7. Select an **Update channel**. The **stable** channel is recommended as it contains the latest stable and *supported* version of the operator.

8. Click the **Install** button:

Install Operator

Install your Operator by subscribing to one of the update channels to keep the Operator up to date. The strategy determines either manual or automatic updates.

Update channel * ⑦

○ preview

◉ stable

Installation mode *

◉ All namespaces on the cluster (default)
  Operator will be available in all Namespaces.

○ A specific namespace on the cluster
  This mode is not supported by this Operator

Installed Namespace *

[PR] openshift-operators ▾

Update approval * ⑦

◉ Automatic

○ Manual

[Install] [Cancel]

Red Hat OpenShift GitOps
provided by Red Hat Inc.
Provided APIs

Ⓐ Application

An Application is a group of Kubernetes resources as defined by a manifest.

AS ApplicationSet

ApplicationSet is the representation of an ApplicationSet controller deployment.

AP AppProject

An AppProject is a logical grouping of Argo CD Applications.

ACD Argo CD

Argo CD is the representation of an Argo CD deployment.

Figure 10.7 – Installing the operator

9. Wait up to 5 minutes until you see the following message:

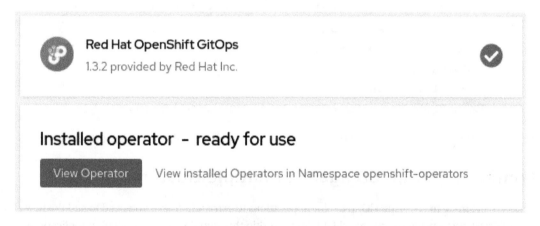

Red Hat OpenShift GitOps
1.3.2 provided by Red Hat Inc.

Installed operator - ready for use

[View Operator] View installed Operators in Namespace openshift-operators

Figure 10.8 – Operator installed

OpenShift GitOps (Argo CD) also has a CLI, which helps execute common tasks, such as updating the admin's password, registering external clusters, and much more. Let's learn how to install the argocd CLI.

Installing the argocd CLI

The `argocd` CLI makes it easier to work with Argo CD. Through it, you can manage Argo CD projects, applications, cluster credentials, and more.

To install the `argocd` CLI, follow these steps:

1. Download the latest Argo CD binary file from `https://github.com/argoproj/argo-cd/releases/latest`.

2. If you are using Linux, download the CLI and add it to your path:

    ```
    $ sudo curl -sSL -o /usr/local/bin/argocd https://github.
    com/argoproj/argo-cd/releases/latest/download/argocd-
    linux-amd64
    $ sudo chmod +x /usr/local/bin/argocd
    ```

3. If everything went well, you will see the following output by running the `argocd version` command. Ignore the error message you see in the last line; it is an expected message as we haven't logged in to any OpenShift cluster yet:

    ```
    $ argocd version
    argocd: v2.2.1+122ecef
       BuildDate: 2021-12-17T01:31:40Z
       GitCommit: 122ecefc3abfe8b691a08d9f3cecf9a170cc8c37
       GitTreeState: clean
       GoVersion: go1.16.11
       Compiler: gc
       Platform: linux/amd64
    FATA[0000] Argo CD server address unspecified
    ```

Now, let's learn how to configure Argo CD to deploy applications against multiple clusters.

Configuring Argo CD against multiple clusters

If you are planning to use Argo CD to deploy applications to *external clusters*, you need to add the new cluster's credentials using the `argocd` CLI. You can skip this step if you want to deploy applications in the same cluster where Argo CD is installed (the `kubernetes.default.svc` file already exists and should be used in this case).

To register new clusters, perform the following steps using the `argocd` CLI you installed previously:

1. Log into the new cluster we want to register:

    ```
    $ oc login -u <user> https://<api-newcluster>:6443
    ```

2. Now, log into the cluster where Argo CD is installed using oc login:

    ```
    $ oc login -u <user> https://<api-argocluster>:6443
    ```

3. At this point, you should have both clusters in your kubeconfig file:

    ```
    $ oc config get-contexts
    CURRENT NAME            CLUSTER     AUTHINFO NAMESPACE
            newcluster      newcluster  admin
    *       argocluster     argocluster
    ```

4. Set a different context for the new cluster:

    ```
    $ oc config set-context prd-cluster --cluster=newcluster
    --user=admin
    Context "prd-cluster" created.
    ```

5. Get the Argo CD public URL from the openshift-gitops namespace:

    ```
    $ oc get route openshift-gitops-server -n openshift-
    gitops -o jsonpath='{.spec.host}'
    ```

6. Get the administrator password:

    ```
    $ oc extract secret/openshift-gitops-cluster -n
    openshift-gitops --to=-
    ```

7. Log in using argocd:

    ```
    $ argocd login --insecure openshift-gitops-server-
    openshift-gitops.apps.example.com
    Username: admin
    Password:
    'admin:login' logged in successfully
    Context ' openshift-gitops-server-openshift-gitops.apps.
    example.com' updated
    ```

8. Now, add the new cluster to Argo CD:

    ```
    argocd cluster add prd-cluster -y
    INFO[0000] ServiceAccount "argocd-manager" created in
    namespace "kube-system"
    INFO[0000] ClusterRole "argocd-manager-role" created
    ```

```
INFO[0001] ClusterRoleBinding "argocd-manager-role-
binding" created
Cluster 'h https://<api-newcluster>:6443' added
```

With that, you are ready to deploy applications either into your local or remote cluster using Argo CD! But before we dive into application deployment, let's check out some of the important aspects related to Argo CD.

Argo CD definitions and challenges

Before we walk through the application deployment process, we need to discuss some important challenges related to GitOps, decisions, and standards.

GitHub repository structure

The first important question that always comes up with GitOps is about the GitHub repository's structure. Should I only use one repository for source code and the Kubernetes manifests? How should I deal with different configuration files for different environments, such as development, QA, and production?

Well, there are no right or wrong answers to these questions, as each option has its pros and cons. You need to find out which approach works best for you. My advice here is: try it! There is nothing better than practical experience, so use each model and find out which one fits best for your applications and teams. In the following sections, we'll look at some of the most popular repository structures out there for GitOps-oriented applications.

Mono-repository

In this structure, you will have one repository for all your Kubernetes manifests and infrastructure-related files. Although there is not a single standard for this structure, you will probably have a repository similar to the following:

```
├── config #[1]
├── environments #[2]
│   ├── dev #[3]
│   │   ├── apps #[4]
│   │   │   └── app-1
│   │   └── env #[5]
│   └── qa #[6]
│       ├── apps
│       │   └── app-1
```

```
|   |      └── env
(...)
```

Let's look at this code in more detail:

- **#[1]**: This folder contains the CI/CD pipelines, Argo CD, and other related configuration files that are common for any environment

- **#[2]**: This folder contains the manifests that are specific to each environment, such as development, QA, and production

- **#[3]**: These are the manifest files that are specific to the development environment

- **#[4]**: Here, you have the Kubernetes manifests to deploy the applications that are tracked and released in this repository

- **#[5]**: These are the cluster or infrastructure-related manifests for the development environment, such as `RoleBinding` permissions, `Namespace`, and so on

- **#[6]**: These are the manifest files that are specific to the QA environment

The main benefit of this approach is its **simplicity**: in this approach, you only need to manage one repository for one or more applications, which makes it easier to manage branches, tags, **PRs**, and anything related to the application's manifests repository. However, the major con of this strategy is that all the contributors can read and make changes to the production manifests. That said, it might be *hard to detect unintentional changes to production*, especially with large PRs.

This leads us to the next approach, in which you have a different repository per environment.

Repository per environment

With this strategy, you will have multiple repositories, one for each environment. In other words, you will have one repository for *development* manifests, another one for *QA*, and so on. In this strategy, you will likely use PRs to promote changes between each environment and have a granular review process, which leads to a less error-prone process. In this strategy, you can also manage Git permissions according to each environment:

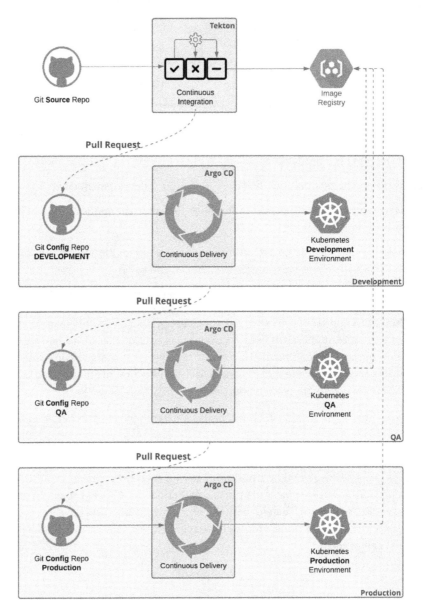

Figure 10.9 – One repository per environment

In this chapter, we will be using a mono-repository strategy and use Git push requests and PRs with multiple branches to mitigate the risk of unintentional changes.

Next, we will discuss another important aspect related to GitOps on Kubernetes: templating YAML files and avoiding duplication.

Templating Kubernetes manifests

Whatever repository structure you decide to go for, one thing is certain: you will need to have separate files and folders for each environment you manage. So, how can you avoid duplicating YAML manifest files everywhere and turning your GitOps process into a nightmare?

Currently, the most popular options to do this are as follows:

- **Helm**: Rely on Helm Charts and Helm Templates to package and deliver Kubernetes applications. Through Helm Templates, you can combine values with templates and generate valid Kubernetes manifest files as a result.

- **Kustomize**: With Kustomize, you can reuse existing manifest files using a patch strategy. It uses a hierarchical structure so that you can flexibly reuse shared configurations and create layers of configurations with only environment-specific parameters that will be overloaded with base parameters.

While Helm is a great package tool, we are going to focus on Kustomize in this chapter due to the following reasons:

- Kustomize runs natively on Kubernetes and the OpenShift CLI (`kubectl/oc`)

- It is declarative, which is an important factor for GitOps, as we mentioned previously

- You can use a remote base in a repository as the starter set of the manifest and have the overlays stored in different repositories

Let's take a closer look at Kustomize.

Kustomize

Kustomize is composed of hierarchical layers of manifest files:

- `Base`: This is a directory that contains the resources that are always reused as the base manifest files. These describe the application and objects declaratively.

- `Overlays`: This is a directory that only contains the configurations that are specific for each overlay. For instance, it is common to have an overlay for the development environment, another for QA, and so on. The configurations that reside in the `overlay` directories replace the values that are in the `base` directory.

You can have multiple layers of bases and overlays – as many as you want. However, to maintain the legibility and maintainability of your application manifest files, it is not recommended to use several layers of manifest files. The following diagram shows an example of a base and two overlays that might be used with Kustomize:

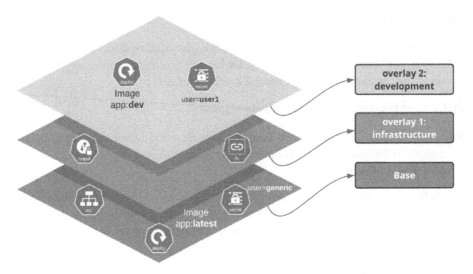

Figure 10.10 – Kustomize layers

The following is a typical folder structure you'll see when you're using Kustomize:

```
├── base
│   ├── deployment.yaml
│   ├── kustomization.yaml
│   ├── route.yaml
│   └── service.yaml
└── overlays
    ├── dev
    │   ├── deployment-patch.yaml
    │   ├── kustomization.yaml
    │   └── namespace.yaml
    ├── prod
    │   ├── deployment-patch.yaml
    │   ├── kustomization.yaml
    │   └── namespace.yaml
    └── qa
        ├── deployment-patch.yaml
        ├── kustomization.yaml
        └── namespace.yaml
```

We will practice using Kustomize a bit more in this chapter when we deploy our example.

Managing secrets

Last, but not least, the real challenge with Kubernetes is managing sensitive data using secrets. While secrets are somewhat safe, depending on how the permissions are set among users, they are not encrypted. This is a real problem when we think about storing those secrets in a GitHub repository. So, how can we handle secrets securely?

There are two ways to handle secrets:

- Use an external vault to store the secrets securely outside Git and the cluster

- Encrypt the secret before saving it in Git using a sealed secret tool such as Bitnami Sealed Secrets

> **Notes**
>
> 1. Secrets are presented in Base64 encoding. The following command, for instance, decrypts a secret named `db-root-password` that contains a password field:
>
> ```
> oc get secret db-root-password -o jsonpath="{.data.password}" | base64 -d
> ```
>
> 2. Bitnami Sealed Secrets allows you to encrypt your secret into a `SealedSecret` object and store it securely, even in a public GitHub repository, since it is encrypted using a public/private certificate. To learn more, check out the link in the *Further reading* section.

With that, we have discussed the main points that you need to think about regarding GitOps. They are important topics we decided to bring to you before our practical example but don't get too worried about that yet – you will find out what works best for you and your team by practicing and learning from it. In the next section, we will introduce some of the main objects you will work with in Argo CD.

Argo CD main objects

In this section, we will look at some of the main Argo CD objects you need to know about. Argo CD is quite simple and most of what you will do can be summarized in two objects: **AppProject** and **Application**.

In this section, we will not mention all the different objects Argo CD has since that is not the main focus of this book. Check out the *Further reading* section to learn more.

AppProject

Projects allow you to group applications and structure them according to any group logic you need. Using projects, you can do the following:

- Limit the Git *source repositories* that can be used to deploy applications

- Restrict the *clusters and namespaces destination* that the applications can be deployed to

- Limit the type of objects that can be deployed (for example, Deployments, Secrets, DaemonSets, and so on)

- Set roles to limit the permissions that are allowed by groups and/or JWTs

When Argo CD is installed, it comes with a `default` project. If you don't specify a project in your Argo CD application, the `default` option will be used. Creating additional projects is optional as you could use Argo CD's `default` project instead. However, it is recommended to create additional projects to help you organize your Argo CD applications.

> **Note**
> There is no relationship between Argo CD's default project and OpenShift's default namespace. Although they have the same name, they are unrelated.

A typical `AppProject` specification looks as follows:

```
apiVersion: argoproj.io/v1alpha1
kind: AppProject
metadata:
  name: clouds-api #[1]
  namespace: openshift-gitops
spec:
  sourceRepos: #[2]
    - '*'
  destinations: #[3]
    - namespace: '*'
      server: '*'
  clusterResourceWhitelist: #[4]
    - group: '*'
      kind: '*'
```

Let's look at this code in more detail:

- **[1]**: The name of the project.

- **[2]**: The Git source repositories that are allowed. In this case, any source repository is allowed

- **[3]**: The destination clusters and namespaces that are allowed. In this case, any combination of clusters and namespaces is allowed

- **[4]**: The objects that can be deployed (for example, Deployments, Secrets, DaemonSets, and so on). In this case, there is no limitation

Properly adjust the code pointed out to achieve manifest file.

Applications

Applications represent application instances that have been deployed and managed by Argo CD. The specification of an application is composed of source and destination. source is where the Kubernetes manifests (Git repository) that specify the desired state of the application reside, while destination specifies the cluster and namespace where the application will be deployed. Besides that, you can also specify the synchronization policies you want Argo CD to apply.

The following is an example of an Application specification:

```
apiVersion: argoproj.io/v1alpha1
kind: Application
metadata:
  name: clouds-app-dev #[1]
  namespace: openshift-gitops #[2]
spec:
  project: clouds-api #[3]
  source: #[4]
    repoURL: `https://github.com/PacktPublishing/OpenShift-
Multi-Cluster-Management-Handbook.git'
    path: chapter10/clouds-api-gitops/overlays/dev
    targetRevision: dev
  destination: #[5]
    server: 'https://kubernetes.default.svc'
    namespace: default
  syncPolicy: #[6]
    automated:
      selfHeal: true
```

Let's look at this code in more detail:

- **[1]**: Argo CD Application name.
- **[2]**: This is the namespace where Argo CD is installed. The default project for the OpenShift GitOps operator is openshift-gitops.
- **[3]**: This is the Argo CD project that you create using the AppProject object. Do not get it confused with the OpenShift project; they are unrelated.
- **[4]**: Git source repository information about where the Kubernetes manifests reside.
- **[5]**: The cluster and namespace where the application will be deployed.
- **[6]**: The synchronization policies that Argo CD will use. We will learn more about these policies in the next section.

> **Important Note**
>
> Argo CD's namespace (`openshift-gitops`) has special privileges within the cluster to perform all the necessary activities. Due to that, you must protect access to this namespace to avoid unwanted deployments or changes.

Syncing policies

You can configure Argo CD to automatically synchronize your application when there is any drift between the desired state specified in the manifests in Git and the actual application state. You have the following options with Argo CD:

- **Self-Healing**: When you set this option to `true`, Argo CD will automatically sync when it detects any differences between the manifests in Git and the actual state. By default, this flag is `false`.

- **Pruning**: As a precaution, automated sync will never delete a resource when it no longer exists in Git. Argo CD only prunes those resources with a manual sync. However, if you want to allow Argo CD to automatically prune objects that no longer exist in Git, you can set the `prune` flag to `true`.

Syncing the order

For standard Kubernetes manifests, Argo CD already knows the correct order that needs to be applied to avoid precedence issues. For instance, consider an application that contains three manifests for namespace creation, deployment, and role bindings. In such a case, Argo CD will always apply the objects in the following order:

1. Namespace
2. Role bindings
3. Deployment

That said, this is the kind of thing you don't need to be worried about as Argo CD is smart enough to apply them in the correct order.

However, there are some other specific cases where you may need to specify objects' precedence. Let's say that you want to deploy an application composed of one StatefulSet to deploy a database and a deployment for an application that uses the database. In this case, you can use **resource hooks** to specify the correct order to apply the objects.

The following types of resource hooks can be used:

- `PreSync`: Objects marked with `PreSync` are executed before any other manifests.

- `Sync`: This runs after `PreSync` is complete. You can also use `sync-wave` to set the sync precedence of the objects in the `Sync` phase.

- `PostSync`: Runs after all the `Sync` objects have been applied and are in a `Healthy` state.

- `SyncFail`: The manifests with this annotation will only be executed when a sync operation fails.

The following is an example of a resource hook specification:

```
apiVersion: apps/v1
kind: Deployment
metadata:
  annotations:
    argocd.argoproj.io/hook: Sync
    argocd.argoproj.io/sync-wave: "1"
(.. omitted ..)
```

There are different types of annotations that you can include in your manifests to perform more complex tasks. Check out the *Further reading* section to learn more.

With that, we have covered the most important concepts and theories behind GitOps and Argo CD. Now, without further ado, let's look at our example and practice what we have discussed so far!

Deploying an application using GitOps

In this practical exercise, we will build and deploy our sample application in three different namespaces to simulate an application life cycle composed of development, QA, and production environments. The following diagram shows the delivery model we will use in this exercise to practice Argo CD deployments. Use it as much as you want as a starting point to build a comprehensive and complex ALM workflow that's suitable for your needs:

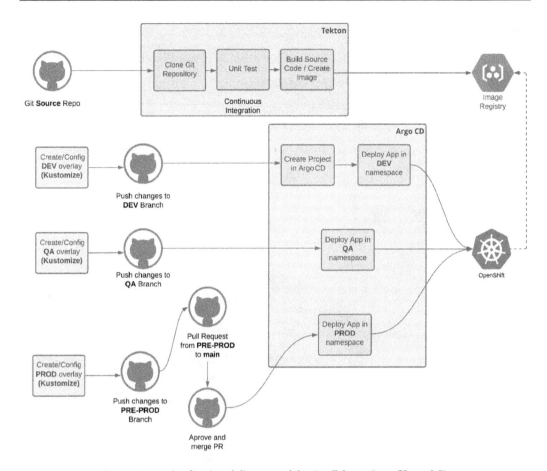

Figure 10.11 – Application delivery model using Tekton, Argo CD, and Git

Once again, we are going to use the content we have prepared in this book's GitHub repository. To do this, you must *fork this repository to your GitHub account*: `https://github.com/ PacktPublishing/OpenShift-Multi-Cluster-Management-Handbook`. Once you have forked it, follow the instructions in this section to put this workflow into practice.

Building a new image version

In this section, we will build a new container image version, 1.0, and push it to the OpenShift internal registry, as shown in the following diagram:

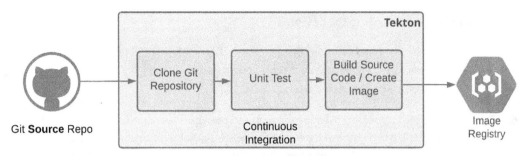

Figure 10.12 – Building a new image version

To do so, perform the following steps:

1. Clone the repository in your machine:

```
$ GITHUB_USER=<your_user>
$ git clone https://github.com/PacktPublishing/OpenShift-
Multi-Cluster-Management-Handbook.git
```

2. Run the following script and follow the instructions to change the references from the original repository (PacktPublishing) to your forked repository:

```
$ cd OpenShift-Multi-Cluster-Management-Handbook/
chapter10
$ ./change-repo-urls.sh
# Go back to the root dir
$ cd ..
```

3. Create a new branch for development:

```
$ git checkout -b dev
```

4. Open the ./sample-go-app/clouds-api/clouds.go file with your preferred text editor and change line 147 by adding version=1.0 to it:

```
$ vim ./sample-go-app/clouds-api/clouds.go
func homePage(w http.ResponseWriter, r *http.Request) {
fmt.Fprintf(w, "Welcome to the HomePage! Version=1.0")
fmt.Println("Endpoint Hit: homePage")
}
```

5. Commit and push change to the dev branch:

```
$ git add ./sample-go-app/clouds-api/clouds.go
$ git commit -m 'Version 1.0 changes'
$ git push -u origin dev
```

6. Run the following command to deploy the required prerequisites and the pipeline that builds image version 1.0, which we will deploy in the development namespace shortly. Make sure that you are already logged into the OpenShift cluster (by using the oc login command):

```
$ oc apply -k ./chapter10/config/cicd
```

7. Now, run the pipeline and check the logs:

```
$ oc apply -f ./chapter10/config/cicd/pipelinerun/build-
v1.yaml -n cicd
$ tkn pipelinerun logs build-v1-pipelinerun -f -n cicd
[fetch-repository : clone] + '[' false = true ']'
[fetch-repository : clone] + '[' false = true ']'
[fetch-repository : clone] + CHECKOUT_DIR=/workspace/
output/
[fetch-repository : clone] + '[' true = true ']'
(.. omitted ..)
[build-image : push] Writing manifest to image
destination
[build-image : push] Storing signatures
 [build-image : digest-to-results] + cat /workspace/
source/image-digest
[build-image : digest-to-results] + tee /tekton/results/
IMAGE_DIGEST
[build-image : digest-to-results] sha256:5c-
c65974414ff904f28f92a0deda96b08f4e-
c5a98a09c59d81eb59459038b547
```

With that, you have built the clouds-api:v1.0 container image and pushed it to OpenShift's internal registry. Now, let's deploy this image using **Kustomize** and **Argo CD**.

Deploying in development

In this section, we are going to use Kustomize to overwrite the image tag of the deployment YAML file so that it uses v1.0, which we built in the previous section. We will also create a new namespace for the development branch named clouds-api-dev.

The following diagram shows the steps we will perform:

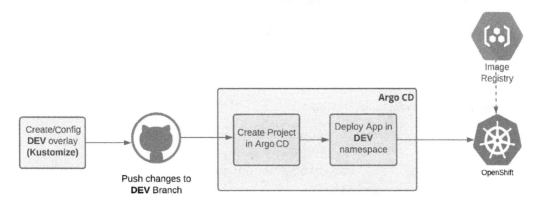

Figure 10.13 – Deploying in development

Perform the following steps:

1. Change the image version of our development kustomization.yaml file. To do so, change line 18 from changeme to v1.0:

```
$ vim ./chapter10/clouds-api-gitops/overlays/dev/
kustomization.yaml
apiVersion: kustomize.config.k8s.io/v1beta1
kind: Kustomization

commonLabels:
  environment: dev

namespace: clouds-api-dev

bases:
  - ../../base

resources:
  - namespace.yaml

images:
- name: quay.io/gfontana/clouds-api
  newName: image-registry.openshift-image-registry.
```

```
svc:5000/cicd/clouds-api
  newTag: v1.0 # Change this line
```

2. Alternatively, you may use the `sed` command to replace this line:

```
sed -i 's/changeme/v1.0/' ./chapter10/clouds-api-gitops/
overlays/dev/kustomization.yaml
```

3. Now, push this change to the `dev` branch:

```
$ git add chapter10/clouds-api-gitops/overlays/dev/
kustomization.yaml
$ git commit -m 'updating kustomization file for v1.0'
$ git push -u origin dev
```

4. Now, let's create a new Argo CD project:

```
$ oc apply -f ./chapter10/config/argocd/argocd-project.
yaml
```

5. Create a new Argo CD application that will deploy the application in the development namespace:

```
$ oc apply -f ./chapter10/config/argocd/argocd-app-dev.
yaml
```

6. Get Argo CD's URL and admin passwords using the following commands:

```
# Get the Argo CD url:
$ echo "$(oc  get route openshift-gitops-server -n
openshift-gitops --template='https://{{.spec.host}}')"
# Get the Admin password
$ oc extract secret/openshift-gitops-cluster -n
openshift-gitops --to=-
```

7. Access the Argo CD UI using the URL and admin user provided previously. You should see a new application there named `clouds-app-dev`:

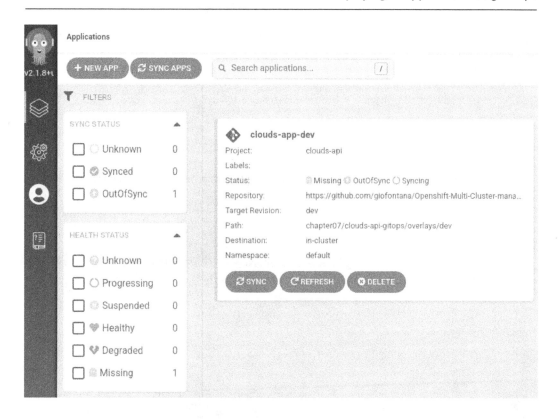

Figure 10.14 – The Argo CD application to deploy in development

8. Click **clouds-app-dev** to learn more about the application:

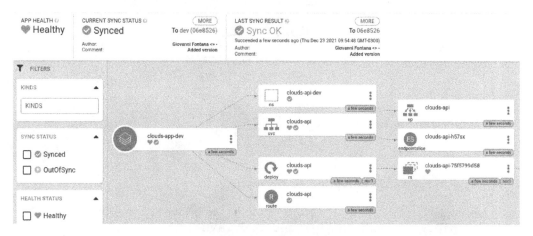

Figure 10.15 – The Argo CD application

9. Run the `curl` command to check that version 1.0 of the application is running and has been successfully deployed by Argo CD:

```
$ curl $(oc get route clouds-api -n clouds-api-dev
--template='http://{{.spec.host}}')
```

You should see the following response:

```
Welcome to the HomePage! Version=1.0
```

With that, our sample application is running in the development namespace. Now, let's learn how to promote this application to the next stage: QA.

Promoting to QA

We have application version 1.0 running in development. Now, let's use Kustomize and Argo CD once more to deploy it in a new namespace that's dedicated to QA, as shown in the following diagram:

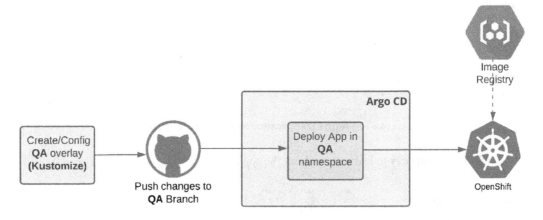

Figure 10.16 – Promoting to QA

Follow these steps:

1. Create a new branch for QA:

```
$ git checkout -b qa
```

2. Create an overlay for QA by copying the `dev` overlay:

```
$ cp -r ./chapter10/clouds-api-gitops/overlays/dev/ ./
chapter10/clouds-api-gitops/overlays/qa/
```

3. Replace the references to dev with qa:

```
$ sed -i 's/dev/qa/' ./chapter10/clouds-api-gitops/
overlays/qa/namespace.yaml ./chapter10/clouds-api-gitops/
overlays/qa/kustomization.yaml
```

4. Push the changes to Git:

```
$ git add ./chapter10/clouds-api-gitops/overlays/qa
$ git commit -m 'Promoting v1.0 to QA'
$ git push -u origin qa
```

5. Deploy the manifest file to promote the environment using Argo CD:

```
$ oc apply -f ./chapter10/config/argocd/argocd-app-qa.
yaml
```

6. Access the Argo CD UI again. At this point, you should have two applications on Argo CD:

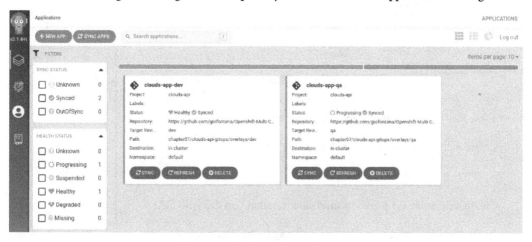

Figure 10.17 – Argo CD applications

7. Let's access the application that is running in the QA namespace:

```
$ curl $(oc get route clouds-api -n clouds-api-qa
--template='http://{{.spec.host}}')
```

You should see the same response that you saw previously:

```
Welcome to the HomePage! Version=1.0
```

With that, we have promoted our application to QA! Now, let's learn how to move it to the last stage, which is the production environment.

Promoting to production

For production, we are going to use a different approach – we are going to use PRs instead of simple Git pushes. We will use a temporary branch named `pre-prod` to commit the overlay manifests that will be used for production, as shown in the following diagram:

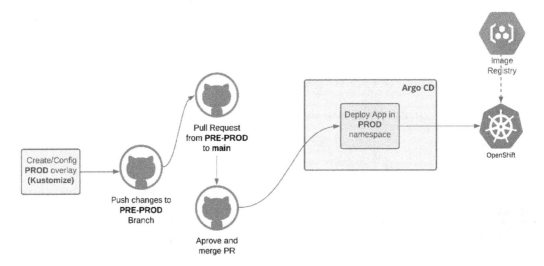

Figure 10.18 – Promoting to production

Follow these steps to promote version 1.0 of our application to production:

1. Create a new branch to prepare for production:

   ```
   $ git checkout -b pre-prod
   ```

2. Create an overlay for production, similar to what you did with QA:

   ```
   $ cp -r chapter10/clouds-api-gitops/overlays/dev/
   chapter10/clouds-api-gitops/overlays/prod/
   $ sed -i 's/dev/prod/' ./chapter10/clouds-api-gitops/
   overlays/prod/namespace.yaml ./chapter10/clouds-api-
   gitops/overlays/prod/kustomization.yaml
   ```

3. Push the changes to the `pre-prod` branch:

   ```
   $ git add ./chapter10/clouds-api-gitops/overlays/prod
   $ git commit -m 'Promoting v1.0 to Prod'
   $ git push -u origin pre-prod
   ```

4. Now, create a PR on GitHub and merge it with the main branch. Access the **Pull requests** tab of your GitHub repository and click the **New pull request** button:

Figure 10.19 – Creating a PR

5. Since you are working in the forked repository, GitHub suggests that you create a PR for the source repository (in this case, from PacktPublishing). We want to create a PR that goes from our pre-prod branch to the main branch, both in our forked repository. So, change the base repository to our forked repository:

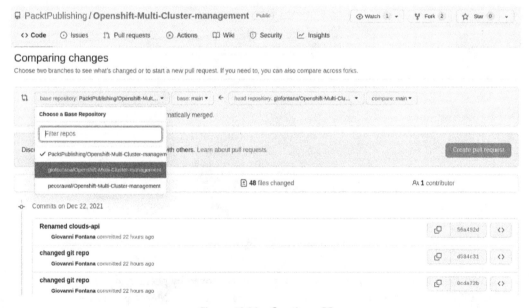

Figure 10.20 – Creating a PR

6. Then, select `pre-prod` in the **compare** field:

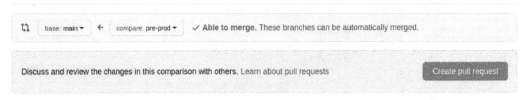

Figure 10.21 – Creating a PR

7. Now, fill out the form and click **Create pull request**:

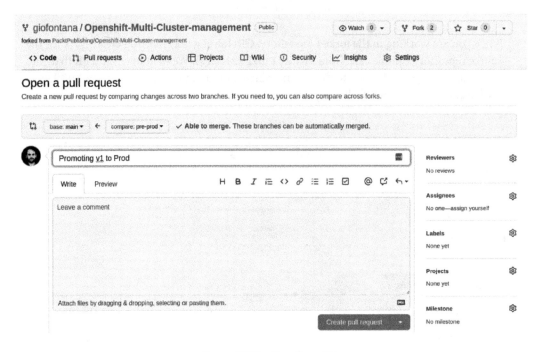

Figure 10.22 – Creating a PR

8. In a real-life scenario, this PR would be reviewed, approved by peers, and then merged. We are still practicing at the moment, so let's go ahead and click the **Merge pull request** button:

Add more commits by pushing to the `pre-prod` branch on **giofontana/Openshift-Multi-Cluster-management**.

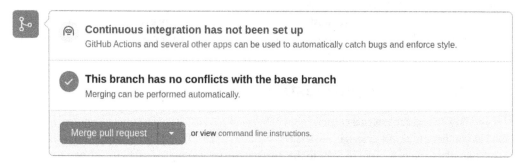

Figure 10.23 – Approving the PR

9. The overlay manifests for version 1.0 of the production environment are already in the `main` branch of our Git repository. This means we can deploy it using Argo CD:

```
$ git checkout main
$ oc apply -f ./chapter10/config/argocd/argocd-app-prod.
yaml
```

10. At this point, you should have three applications on Argo CD:

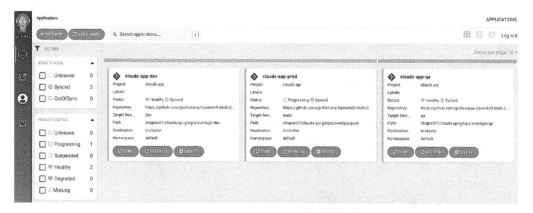

Figure 10.24 – Argo CD applications

11. Let's access the application to see version 1.0 of our application running in production:

```
$ curl $(oc get route clouds-api -n clouds-api-prod
--template='http://{{.spec.host}}')
```

You should see the same response that you saw previously:

```
Welcome to the HomePage! Version=1.0
```

Congratulations! We have deployed our application using Argo CD into three different namespaces, each one representing a different environment: development, QA, and production. Since this book is intended to be about *multi-cluster*, we must learn how to do the same process but deploy into multiple clusters, instead of only one. In the next section, you will see that the process is the same, except you must change one parameter in Argo CD's `Application` object.

Deploying to multiple clusters

We learned how to register external clusters in the *Configuring Argo CD against multiple clusters* section. As soon as you have multiple external clusters registered to Argo CD, deploying an application to one of them is simple – you only need to refer to the external cluster you registered in the `destination` field of Argo CD's `Application`. An example of this can be seen in the following manifest:

```
apiVersion: argoproj.io/v1alpha1
kind: Application
metadata:
  name: clouds-app-dev-external-cluster
  namespace: openshift-gitops
spec:
  project: clouds-api
  source:
    repoURL: `https://github.com/PacktPublishing/OpenShift-
Multi-Cluster-Management-Handbook'
    path: chapter10/clouds-api-gitops/overlays/dev
    targetRevision: dev
  destination:
    server: 'https://api.<external-cluster>:6443'
    namespace: default
  syncPolicy:
    automated:
      selfHeal: true
```

You can create as many `Application` objects as you need, deploying only to the local cluster or including multiple external clusters. As you have seen, the deployment process itself is similar, regardless of whether you are deploying to a local or external cluster.

> **Important Note**
>
> When you work with multiple clusters, you need to pay special attention to the **container image registry**. The OpenShift internal registry, as its name suggests, should only be used internally in a single cluster; it is not suitable for multiple clusters. In such a case, an enterprise container image registry is recommended. There are multiple options on the market, such as Nexus, Quay, Harbor, and many others. In this book, we will cover Quay in *Chapter 13, OpenShift Plus – a Multi-Cluster Enterprise-Ready Solution.*

Summary

In this chapter, you learned about various concepts related to **GitOps**. You also learned about **Argo CD** and how to install it on OpenShift and use it. You also built and deployed a sample application to three different namespaces to simulate the *development*, *QA*, and *production* environments. Finally, you learned that deploying to the local or external cluster is a similar process – you only need to change the destination server field.

Argo CD allows you to establish an efficient and robust application delivery model using GitOps, in which you ensure *consistency*, *auditable changes*, and a *secure process*, no matter where you are deploying your applications. And the best part is that there is no additional cost to use it since it is included in Red Hat OpenShift's subscription. That said, if you are deploying containerized applications on OpenShift, I strongly recommend that you try OpenShift GitOps and use the concepts we explored in this chapter.

In the next chapter, we will explore a great tool that will help you deploy and manage several OpenShift clusters from a single unified interface – **Red Hat Advanced Cluster Management**. This tool allows you to monitor, manage, define, and enforce policies and deploy applications to several clusters.

Let's move on and take a deep dive into Red Hat Advanced Cluster Management!

Further reading

To find out more about the topics that were covered in this chapter, take a look at the following resources:

- *A History of GitOps:* `https://www.weave.works/blog/the-history-of-gitops`

- *Argo CD official documentation:* `https://argo-cd.readthedocs.io/`

- *Great tutorial about* **Kustomize**: `https://blog.stack-labs.com/code/kustomize-101/`

- *Bitnami's Sealed Secrets overview:* `https://github.com/bitnami-labs/sealed-secret`.

11
OpenShift Multi-Cluster GitOps and Management

In the first chapter of this book, we discussed the main challenges most organizations face in scaling their Kubernetes infrastructure in a multi- or hybrid-cloud world. New challenges arise when you deploy multiple clusters on different providers, such as the following:

- **Inconsistent security policies**: Clusters with different configurations regarding access rules and user profiles, allowed/blocked network flows, certificates, and other security aspects make an organization more vulnerable to data breaches and other security events.

- **High operational effort to manage all clusters**: Managing configurations, components, compliance, and policies for several clusters is overwhelming.

- **Deploying and managing applications**: The deployment process becomes much more complicated when you need to do it over several clusters. Monitoring and managing them are also really complex and require a lot of human effort.

In this chapter, we will introduce a great tool to help you address these challenges and alleviate the amount of work you and/or your team may need to deal with when managing several clusters: Red Hat **Advanced Cluster Management** (**ACM**).

Therefore, you will find the following topics covered in this chapter:

- What is Red Hat ACM?

- Red Hat ACM installation

- Managing clusters using Red Hat ACM

- Managing applications using Red Hat ACM

- Governance using Red Hat ACM

- Multi-cluster observability with Red Hat ACM

Note

The source code used in this chapter is available at `https://github.com/ PacktPublishing/OpenShift-Multi-Cluster-Management-Handbook/ tree/main/chapter11`.

What is Red Hat ACM?

Red Hat ACM is a complete solution for Kubernetes multi-cluster management from a single pane that includes some other great features, making complex and time-consuming tasks a lot easier. Red Hat ACM provides a few main features, listed here:

- **Kubernetes multi-cluster management**: Create, update, and delete Kubernetes clusters on-premises and in the cloud.

- **Multi-cluster observability**: ACM can also provide observability for all clusters from a single point of view, enabling administrators to read, aggregate, and receive alerts on the clusters.

- **Governance using policies**: Through Red Hat ACM, you can audit and enforce policies to apply anything you want in the clusters, from security to infrastructure and application-related stuff. This includes roles and access control, operators that must be installed, and security compliance rules.

- **Application management**: Deploy applications from Git repositories or Helm into multiple clusters simultaneously and also view them from a single pane.

One of the great aspects of ACM is its multi-cluster architecture – it is designed to manage several clusters from a single standpoint, as you can see in the following figure.

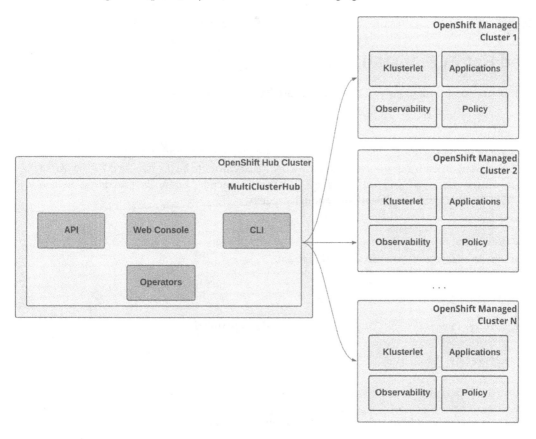

Figure 11.1 – ACM hub and managed clusters

To do so, it uses the concept of hub and managed clusters, as follows:

- **Hub cluster**: An OpenShift cluster that runs the central controller of ACM, which contains the web console, APIs, and other components that make up the product's features. In this chapter, we will use the term hub cluster several times to refer to the OpenShift cluster that hosts ACM. OpenShift is the only supported option for the hub cluster.

- **Managed cluster**: A cluster that is managed by ACM. ACM can manage OpenShift as well as other Kubernetes-based distributions. Check the *Further reading* section at the end of this chapter to find a complete list of Kubernetes distributions supported by ACM (Supportability Matrix).

We will dig into all these features in the following sections of this chapter.

Red Hat ACM installation

In this section, we will guide you through the installation and configuration of ACM.

> **Important Note**
> It is important to consider that ACM uses the compute, memory, and storage resources of the hub cluster, thus it is recommended to have a dedicated cluster to be the hub for ACM, avoiding concurrent workloads and resource usage. This is recommended but not required; you can run ACM in any OpenShift cluster with enough capacity.

The installation process is simple, similar to what we followed in the last chapters with OpenShift Pipelines and GitOps, as you can see in this section.

Prerequisites

1. Access to an OpenShift cluster with cluster-admin permissions.

Installation

Follow this process to install Red Hat Advanced Cluster Management:

1. Access the OpenShift web console using a cluster-admin user.
2. Navigate to the **Operators | OperatorHub** menu item.

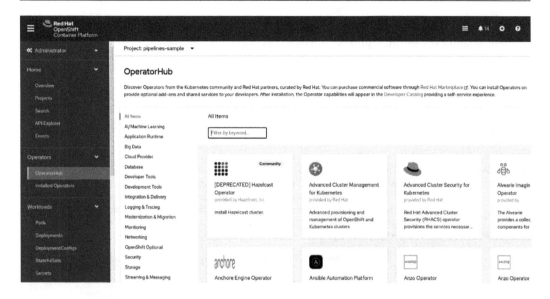

Figure 11.2 – OperatorHub

3. Search for `Advanced Cluster Management for Kubernetes` using the **Filter by keyword…** box.

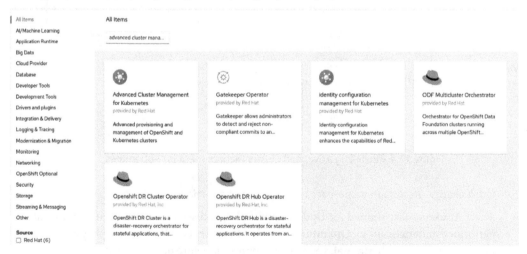

Figure 11.3 – Advanced Cluster Management for Kubernetes on OperatorHub

4. Click on the **Advanced Cluster Management for Kubernetes** tile and then on the **Install** button to see the **Install Operator** screen.

 ## Advanced Cluster Management for Kubernetes
2.5.0 provided by Red Hat

`Install`

Latest version

2.5.0

Capability level

✔ Basic Install
✔ Seamless Upgrades
○ Full Lifecycle
○ Deep Insights
○ Auto Pilot

Source

Red Hat

Provider

Red Hat

Infrastructure features

Disconnected
Proxy-aware
FIPS Mode

Valid Subscriptions

OpenShift Platform Plus
Red Hat Advanced Cluster
Management for
Kubernetes

Red Hat Advanced Cluster Management for Kubernetes provides the multicluster hub, a central management console for managing multiple Kubernetes-based clusters across data centers, public clouds, and private clouds. You can use the hub to create Red Hat OpenShift Container Platform clusters on selected providers, or import existing Kubernetes-based clusters. After the clusters are managed, you can set compliance requirements to ensure that the clusters maintain the specified security requirements. You can also deploy business applications across your clusters.

Red Hat Advanced Cluster Management for Kubernetes also provides the following operators:

- Multicluster subscriptions: An operator that provides application management capabilities including subscribing to resources from a channel and deploying those resources on MCH-managed Kubernetes clusters based on placement rules.
- Hive for Red Hat OpenShift: An operator that provides APIs for provisioning and performing initial configuration of OpenShift clusters. These operators are used by the multicluster hub to provide its provisioning and application-management capabilities.

How to Install

Use of this Red Hat product requires a licensing and subscription agreement.

Install the Red Hat Advanced Cluster Management for Kubernetes operator by following instructions presented when you click the `Install` button. After installing the operator, create an instance of the `MultiClusterHub` resource to install the hub. Note that if you will be using the hub to manage non-OpenShift 4.x clusters, you will need to create a Kubernetes `Secret` resource containing your OpenShift pull secret and specify this `Secret` in the `MultiClusterHub` resource, as described in the install documentation.

You can find additional installation guidance in the install documentation.

Figure 11.4 – Installing Advanced Cluster Management for Kubernetes

5. Don't change the default namespace (`open-cluster-management`).

6. Select **Automatic** or **Manual** for **Update approval**. If you select **Automatic**, upgrades will be performed automatically by **Operator Lifecycle Manager** (**OLM**) as soon as they are released by Red Hat, while with **Manual**, you need to approve it before being applied.

7. Select the correct update channel. The stable channel is recommended as it contains the latest stable and *supported* version of the operator.

8.   Click the **Install** button.

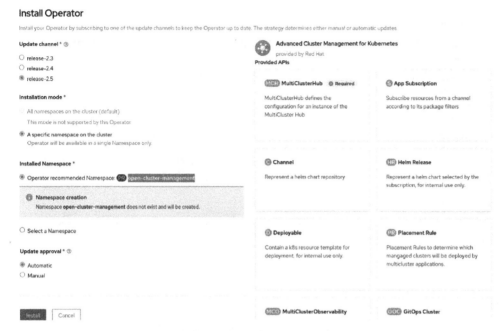

Figure 11.5 – Installing the operator

9.   Wait up to 5 minutes until you see the following message:

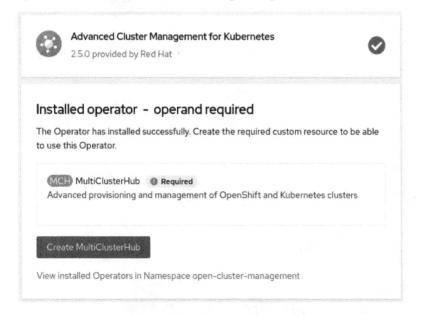

Figure 11.6 – Operator installed

Now that we have the operator installed, we can go ahead and deploy a new **MultiClusterHub** instance:

1. Click on the **Create MultiClusterHub** button.
2. Usually, no changes are needed; leave the default values and click on the **Create** button. Check out the link in the *Further reading* section of this chapter for product documentation for more information if you need to configure some advanced settings.

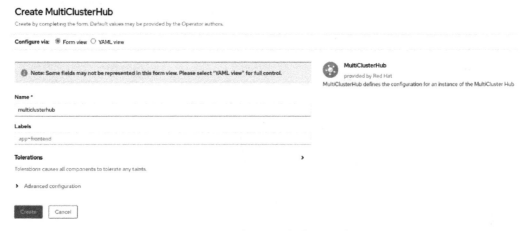

Figure 11.7 – Create MultiClusterHub

3. You will see the status **Phase:Installing** during the installation process.

Figure 11.8 – Installing MultiClusterHub

4. Wait a bit until you see the **Phase:Running** status.

Figure 11.9 – MultiClusterHub running

5. Click on the *combo box* at the top left of the screen and then click on **Advanced Cluster Management**.

Figure 11.10 – Advanced Cluster Management option

6. You should see the ACM login screen. Use the same admin credentials you use to log in to OpenShift.

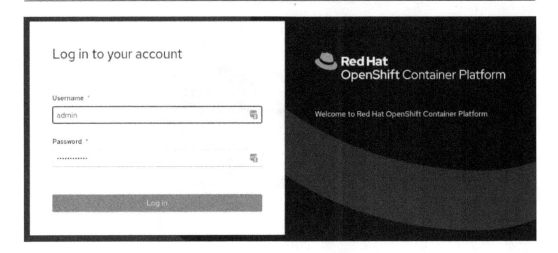

Figure 11.11 – Red Hat ACM login

You have now Red Hat ACM installed and ready to be used.

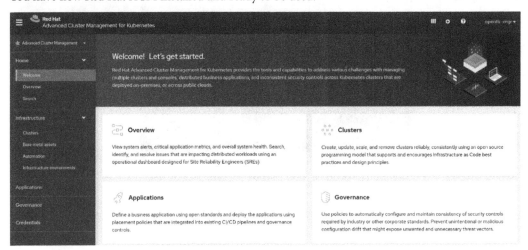

Figure 11.12 – Red Hat ACM initial page

Continue to the next section to learn more about the ACM cluster management feature.

Managing clusters using Red Hat ACM

As we mentioned previously, one of the features ACM provides is cluster management. The following is a list of some of the operations you can perform with ACM:

- Cluster provisioning

- Import of existing clusters

- Destroying a cluster

- Upgrading a cluster

- Scaling cluster nodes in or out

Check the *Further reading* section of this chapter to see a link to a complete list of supported operations.

We will not cover how to do all the operations you can perform with ACM in this book, but we will guide you through the process of provisioning a new OpenShift cluster on AWS using ACM, to give you an idea of how easy it is to use the tool.

Cluster provisioning

Currently, in version 2.5, ACM can deploy clusters on AWS, Azure, Google Cloud, VMware vSphere, bare metal, Red Hat OpenStack, and Red Hat Virtualization. To do so, you need to first input the provider credentials to be used by ACM during the provisioning process. The following steps show how to add AWS credentials that will be used with our sample:

1. Access the **Credentials** menu and click on the **Add credential** button.

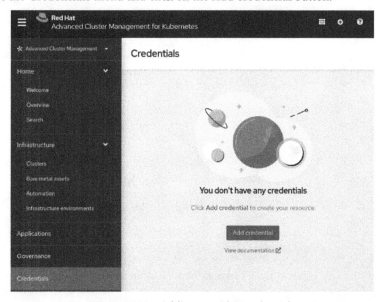

Figure 11.13 – Adding provider credentials

2. Click on the AWS credential.

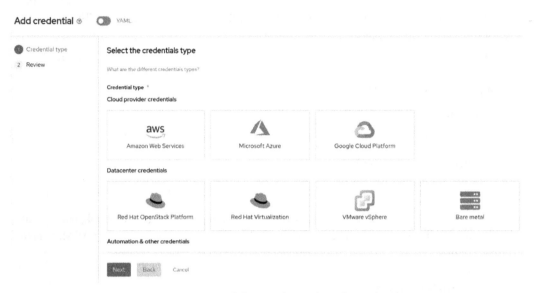

Figure 11.14 – Selecting the credential type

3. On the next page, input a name and select a namespace where the credentials will be stored and the base DNS domain that will be used to deploy clusters. Click on the **Next** button.

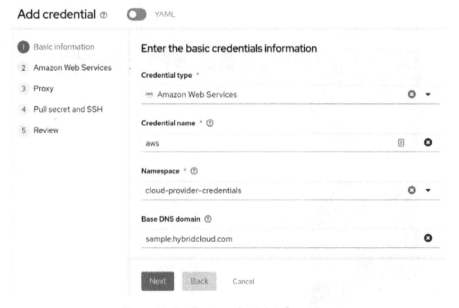

Figure 11.15 – Basic credentials information

Recommended Practice

The provider credentials are stored in secrets in the namespace provided. As such, it is highly recommended you create a specific namespace for that and keep the access for it restricted.

4. Enter the AWS access and secret keys. Click **Next**.

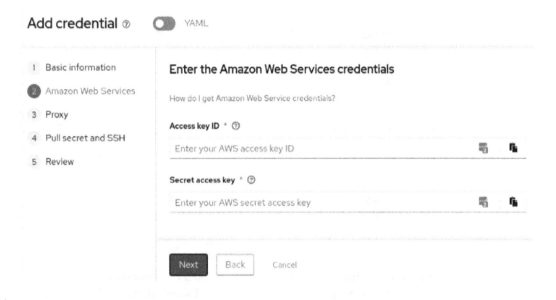

Figure 11.16 – AWS access and secret keys

5. If you use a proxy, input the proxy configuration and click **Next**.

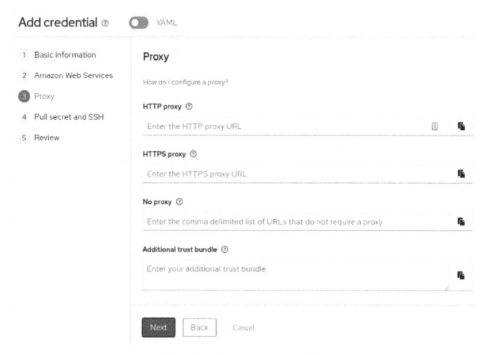

Figure 11.17 – Proxy configuration

6. Log in to the `console.redhat.com` portal using your Red Hat credentials and go to the **OpenShift** | **Downloads** menu option. Scroll down in the **Tokens** section and click on the **Copy** button next to **Pull Secret**.

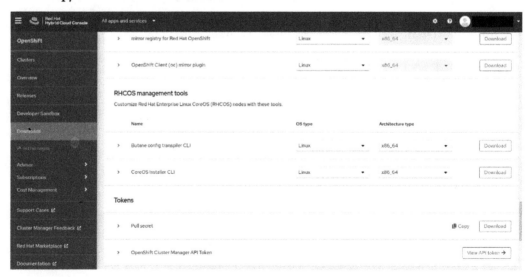

Figure 11.18 – Getting a pull secret

7. Go back to ACM and paste the pull secret in the required field. Use an existing or create a new SSH key and paste it into the **SSH private key** and **SSH public key** fields.

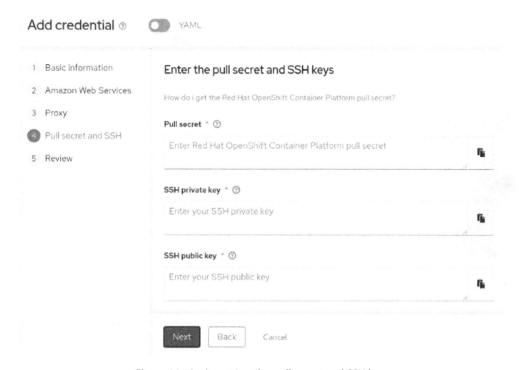

Figure 11.19 – Inputting the pull secret and SSH keys

> **Note**
>
> You can use the following command in a Linux workstation to generate a new SSH key if needed:
>
> ```
> ssh-keygen -t ed25519 -N '' -f new-ssh-key
> ```

8. Finally, click on the **Add** button and see your new credential.

Credentials

Figure 11.20 – Credential added

Now, let's go ahead and deploy a new cluster using this credential. Follow this process to deploy the cluster using ACM:

1.   Access the **Infrastructure | Clusters** menu and click on the **Create cluster** button.

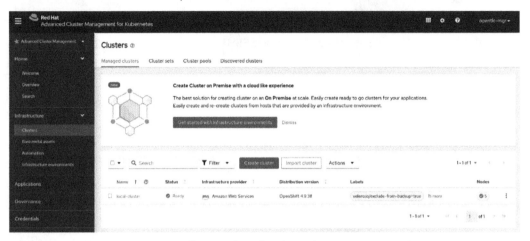

Figure 11.21 – Creating a cluster

2.   Select AWS, choose the credential that we just created, and click **Next**.

Figure 11.22 – Selecting the installation type

3. Fill out the form with the requested fields and click on the **Next** button.

Figure 11.23 – Filling out the cluster details

4. Input the AWS region and the machine number and size.

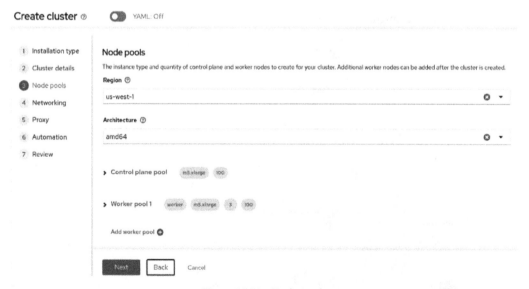

Figure 11.24 – Node pools

5. If you want to customize the cluster network configurations, you can do so on this screen.

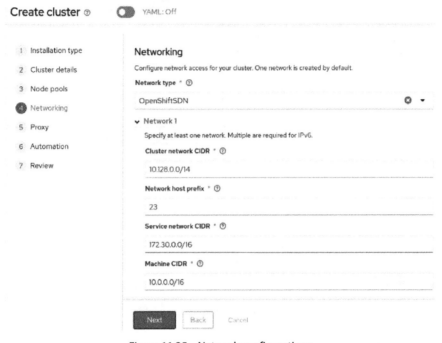

Figure 11.25 – Network configurations

6. Input the proxy configurations, if needed.

Figure 11.26 – Proxy configuration

7. You can also use Ansible to run playbooks and automate infrastructure requirements you may have as part of your deployment process. We will not dig into the Ansible integration in this book, but you can find good references to it in the *Further reading* of this chapter.

Figure 11.27 – Ansible automation hooks

8. Review the information provided and click on the **Create** button if everything is correct.

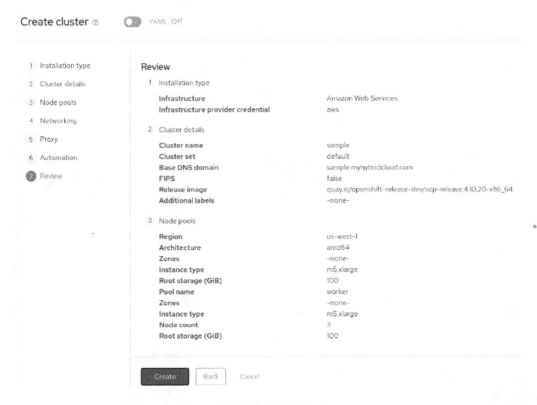

Figure 11.28 – Reviewing a cluster

9. You will be redirected to the overview page, where you can inspect the installation process. The deployment process usually takes between 30 and 60 minutes depending on the provider and region.

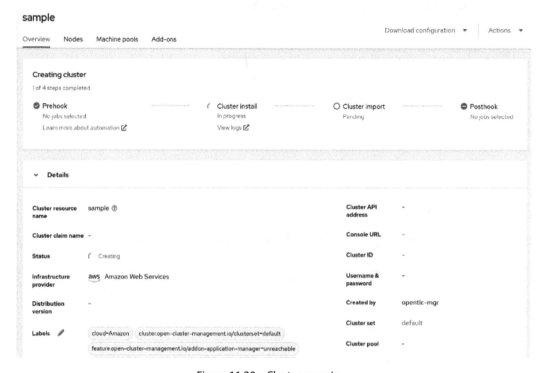

Figure 11.29 – Cluster overview

10. It is recommended that you add labels in your clusters according to some organizational structure to be used later with **PlacementRules**. We are going to add the env=dev label in this cluster and use it in the next section when we deploy an application into a remote cluster using ACM. To do so, click on the pencil icon next to the **Labels** section.

sample

Overview   Nodes   Machine pools   Add-ons

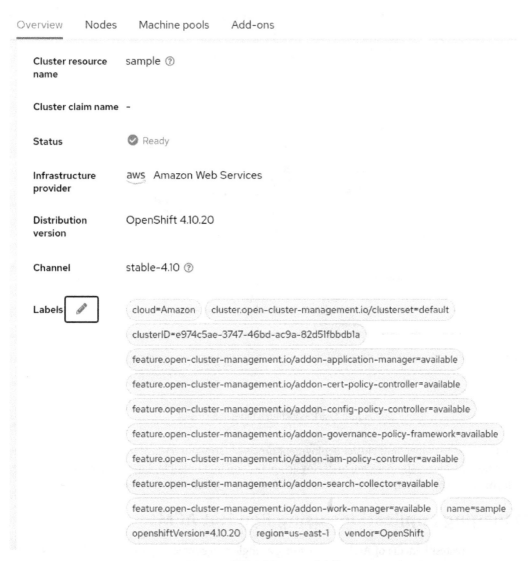

Cluster resource name	sample ⓘ
Cluster claim name	–
Status	✓ Ready
Infrastructure provider	aws Amazon Web Services
Distribution version	OpenShift 4.10.20
Channel	stable-4.10 ⓘ
Labels ✏️	cloud=Amazon · cluster.open-cluster-management.io/clusterset=default
	clusterID=e974c5ae-3747-46bd-ac9a-82d51fbbdb1a
	feature.open-cluster-management.io/addon-application-manager=available
	feature.open-cluster-management.io/addon-cert-policy-controller=available
	feature.open-cluster-management.io/addon-config-policy-controller=available
	feature.open-cluster-management.io/addon-governance-policy-framework=available
	feature.open-cluster-management.io/addon-iam-policy-controller=available
	feature.open-cluster-management.io/addon-search-collector=available
	feature.open-cluster-management.io/addon-work-manager=available · name=sample
	openshiftVersion=4.10.20 · region=us-east-1 · vendor=OpenShift

Figure 11.30 – Adding new labels

11. Then, add the new label, `env=dev`.

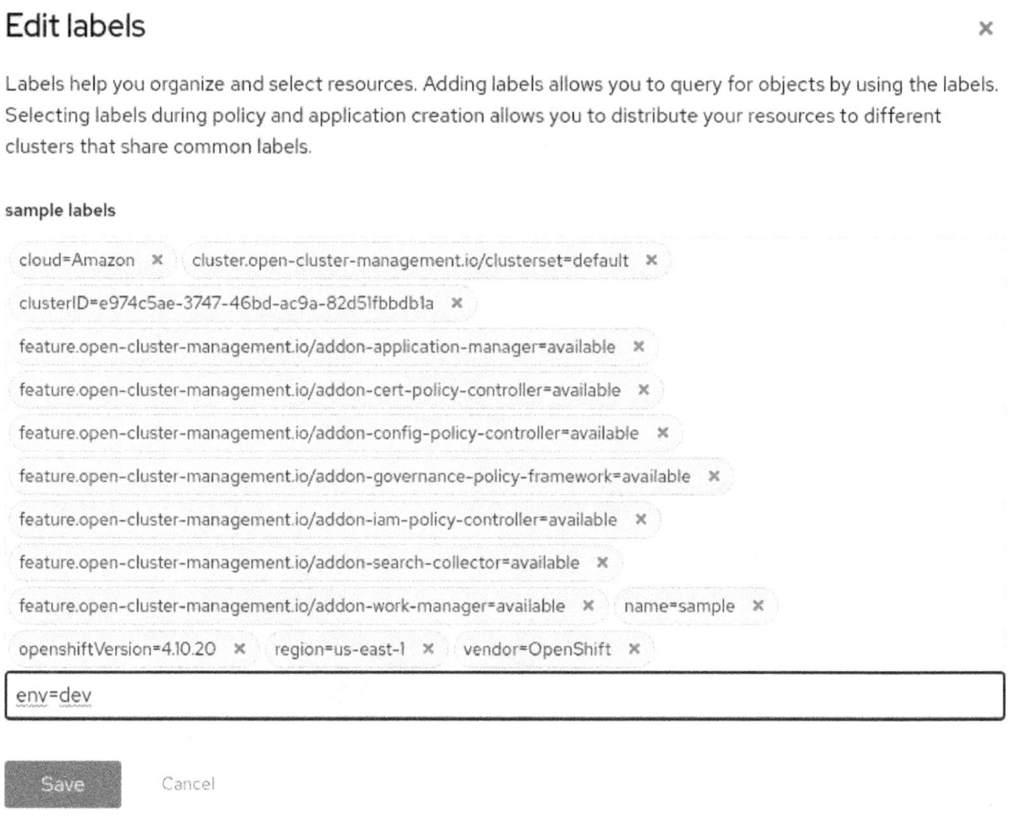

Edit labels   ✕

Labels help you organize and select resources. Adding labels allows you to query for objects by using the labels.
Selecting labels during policy and application creation allows you to distribute your resources to different
clusters that share common labels.

sample labels

cloud=Amazon ✕   cluster.open-cluster-management.io/clusterset=default ✕

clusterID=e974c5ae-3747-46bd-ac9a-82d51fbbdb1a ✕

feature.open-cluster-management.io/addon-application-manager=available ✕

feature.open-cluster-management.io/addon-cert-policy-controller=available ✕

feature.open-cluster-management.io/addon-config-policy-controller=available ✕

feature.open-cluster-management.io/addon-governance-policy-framework=available ✕

feature.open-cluster-management.io/addon-iam-policy-controller=available ✕

feature.open-cluster-management.io/addon-search-collector=available ✕

feature.open-cluster-management.io/addon-work-manager=available ✕   name=sample ✕

openshiftVersion=4.10.20 ✕   region=us-east-1 ✕   vendor=OpenShift ✕

env=dev

Save   Cancel

Figure 11.31 – Adding a label in a cluster

As you can see, the OpenShift cluster deployment process is straightforward! In the next section, you
will see how ACM can also help you to deploy an application into multiple clusters using its embedded
deployment mechanism or also using it integrated with OpenShift GitOps (Argo CD).

Managing applications using Red Hat ACM

One of the greatest benefits of ACM is providing a single and simple way to view applications that
are deployed among different clusters. You can also deploy an application into multiple clusters using
two different approaches:

- Using the embedded Application Subscription deployment model
- Using OpenShift GitOps (Argo CD) and **ApplicationSets**

We will walk through the process of each approach in this section.

Application Subscription model

This model is embedded in ACM and doesn't depend on anything other than ACM itself. In the Application Subscription model, you will define an **Application** object that subscribes (**Subscription**) to one or more Kubernetes resources (**Channel**) that contain the manifests that describe how the application is deployed. The application will be deployed in the clusters defined in the placement rules.

The following is a diagram that explains how this model works:

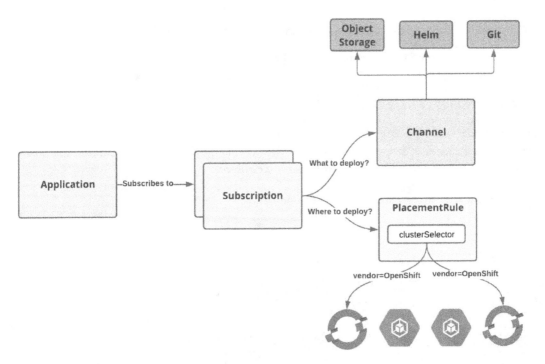

Figure 11.32 – ACM Application Subscription model

Let's get back to the sample application we used in the previous chapter and create the ACM objects to check what the application deployment model looks like.

Channels (what to deploy)

Define the source repositories used to deploy an application. It can be a Git repository, Helm release, or object storage repository. We are going to use the following YAML manifest to point out to our Git repository:

```
apiVersion: apps.open-cluster-management.io/v1
kind: Channel
```

```
metadata:
  name: cloud-api-github
  namespace: clouds-api-dev
spec:
  pathname: https://github.com/PacktPublishing/OpenShift-Multi-
Cluster-Management-Handbook.git #[1]
  type: Git
```

#[1] highlights the URL to the Git repository that contains the application deployment manifests.

After the Channel object, we need to create the PlacementRule object, which will be used with the application deployment.

Placement rules (where to deploy)

Placement rules define the target clusters where the application will be deployed. They are also used with policies. Remember that we added the env=dev label to the cluster we provisioned earlier. We are going to use it now to define our PlacementRule object:

```
apiVersion: apps.open-cluster-management.io/v1
kind: PlacementRule
metadata:
  name: cloud-api-placement
  namespace: clouds-api-dev
  labels:
    app: cloud-api
spec:
  clusterSelector:
    matchLabels:
      env: dev #[1]
```

#[1] highlights the cluster selector based on labels. It will instruct ACM to deploy the application in all clusters that have the env=dev label.

We are now ready to create the Subscription object.

Subscriptions

Subscriptions are used to subscribe clusters to a source repository and also define where the application will be deployed. They work like a glue between the deployment manifests (Channel) and the target clusters (PlacementRule). The following shows what our Subscription object looks like:

```
apiVersion: apps.open-cluster-management.io/v1
kind: Subscription
metadata:
  name: cloud-api-subscription
  namespace: clouds-api-dev
  annotations:
    apps.open-cluster-management.io/git-path: sample-go-app/
clouds-api/k8s/ #[1]
  labels:
    app: cloud-api #[2]
spec:
  channel: clouds-api-dev/cloud-api-github #[3]
  placement:
    placementRef: #[4]
      name: cloud-api-placement
      kind: PlacementRule
```

In the preceding code, we have highlighted some parts with numbers. Let's take a look:

- **#[1]**: Path for deployment manifests on Git.
- **#[2]**: Subscription labels. Used later with the Application object.
- **#[3]**: Channel that contains the Git repository, Helm, or object storage.
- **#[4]**: PlacementRule, which describes where the application will be deployed.

Finally, we can now create the ACM Application object.

Applications

Applications are objects used to describe a group of ACM resources that are needed to deploy an application. The following is the Application object of our sample:

```
apiVersion: app.k8s.io/v1beta1
kind: Application
metadata:
```

```
    name: cloud-api
    namespace: clouds-api-dev
spec:
  componentKinds:
    - group: apps.open-cluster-management.io
      kind: Subscription
  descriptor: {}
  selector:
    matchExpressions: #[1]
      - key: app
        operator: In
        values:
          - cloud-api
```

#[1] highlights subscriptions used by this application. In this case, subscriptions that have the app=cloud-api label will be used.

Now that we understand the objects involved in application deployment, let's create them on ACM.

Deploying the application

Deploying the objects is as simple as running an oc apply command from the hub cluster. Run the following commands from the hub cluster to deploy the application:

```
$ git clone https://github.com/PacktPublishing/OpenShift-Multi-
Cluster-Management-Handbook.git
$ cd OpenShift-Multi-Cluster-Management-Handbook/
$ oc apply -k chapter11/acm-model
namespace/clouds-api-dev created
application.app.k8s.io/cloud-api created
channel.apps.open-cluster-management.io/cloud-api-github
created
placementrule.apps.open-cluster-management.io/cloud-api-
placement created
subscription.apps.open-cluster-management.io/cloud-api-
subscription created
```

You can check the application status by running the following command:

```
$ oc get application -n clouds-api-dev
NAME          TYPE    VERSION   OWNER   READY   AGE
cloud-api                                       5m48s
```

You can alternatively deploy the application using the ACM web console. To do so, perform the following process:

1. Access the **Applications** | **Create application** | **Subscription** menu option.

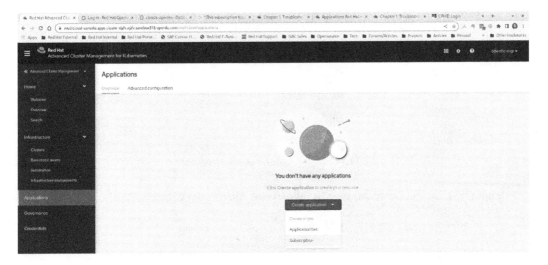

Figure 11.33 – Deploying an application using ACM

2. Fill out the form with the application data, including **Name** and **Namespace**, choose the **Git** repository, and then click on the **Create** button.

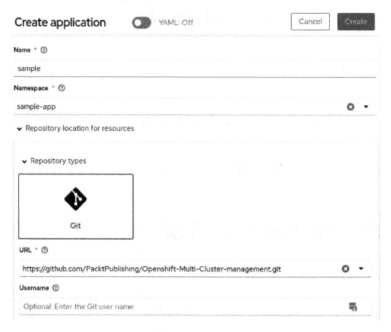

Figure 11.34 – Filling out the application data

3.   Input the placement configuration as follows and click on the Create button.

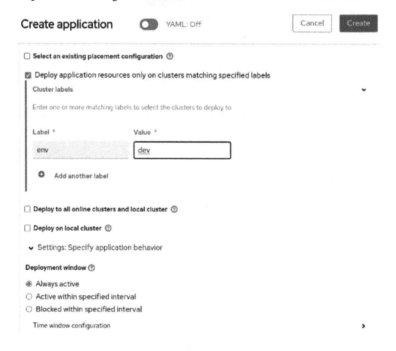

Figure 11.35 – Placement configuration details

4.   Click on the **Topology** tab to see an overview of the application deployed.

sample

Actions   ▼

Overview   Topology

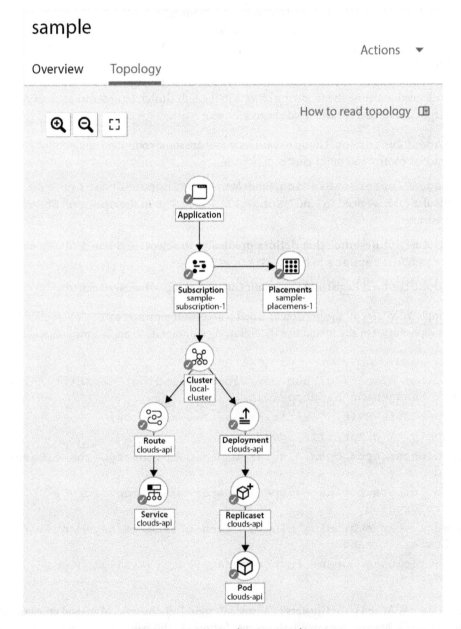

Figure 11.36 – Application topology

Now that you know how to deploy an application using the embedded ACM subscription, let's see how we would do the same using OpenShift GitOps (Argo CD).

OpenShift GitOps (Argo CD) and ApplicationSets

As already mentioned, you can alternatively deploy applications using ACM integrated with OpenShift GitOps (Argo CD) through an object called an **ApplicationSet**. In the *Configuring Argo CD for multi-cluster* section of *Chapter 10, OpenShift GitOps – Argo CD*, we saw how to use the `argocd` command-line option to add managed clusters to Argo CD. You don't need to do that when you use ACM, as ACM manages the external clusters and can add them to Argo CD for you. Instead, with ACM you will need to define the following objects in the hub cluster, in order to instruct ACM to configure Argo CD and add the managed clusters for you:

- `ManagedClusterSet`: Group of clusters that share some common configurations, such as user access control and multi-cluster networking.

- `ManagedClusterSetBinding`: Binds `ManagedClusterSet` to a namespace. In this particular case, we need to bind `ManagedClusterSet` to the `openshift-gitops` namespace.

- `Placement`: Resource that defines predicates to select `ManagedCluster` from `ManagedClusterSets` to bind to `GitOpsCluster`.

- `GitOpsCluster`: Register to OpenShift GitOps a group of managed clusters.

We have sample YAML for all the previously listed objects in the `chapter11/argocd` folder of our GitHub repository. Go ahead and use the following commands to apply those objects in your hub cluster:

```
$ git clone https://github.com/PacktPublishing/OpenShift-Multi-
Cluster-Management-Handbook.git
$ cd OpenShift-Multi-Cluster-Management-Handbook/
$ oc apply -k chapter11/argocd
gitopscluster.apps.open-cluster-management.io/argo-acm-clusters
created
managedclusterset.cluster.open-cluster-management.io/all-
clusters created
managedclustersetbinding.cluster.open-cluster-management.io/
all-clusters created
placement.cluster.open-cluster-management.io/all-clusters
created
```

Now, access your ACM and go to **Clusters | Cluster sets** (tab) | **all-clusters** | **Managed clusters** (tab), and then click on the **Manage resource assignments** button. On this page, select all your clusters and click on the **Review** button and then **Save**.

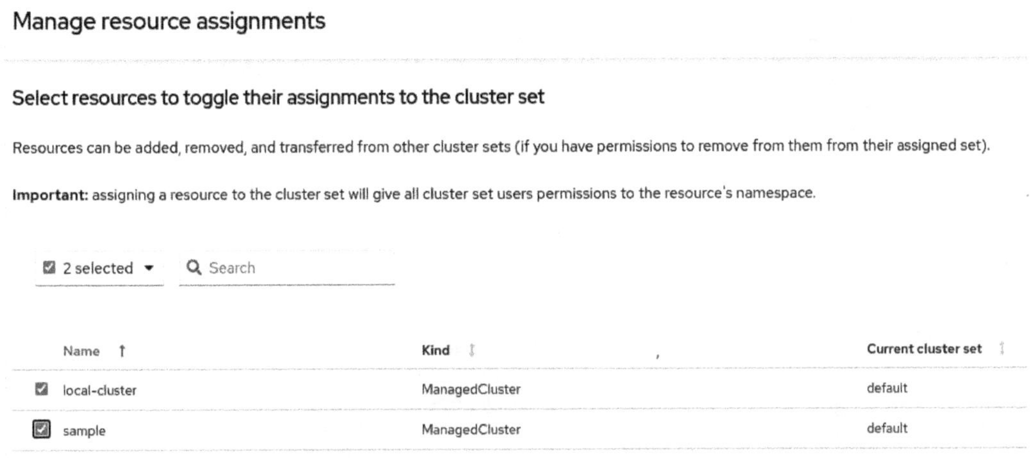

Manage resource assignments

Select resources to toggle their assignments to the cluster set

Resources can be added, removed, and transferred from other cluster sets (if you have permissions to remove from them from their assigned set).

Important: assigning a resource to the cluster set will give all cluster set users permissions to the resource's namespace.

☑ 2 selected ▾   🔍 Search

Name ↑	Kind ↕		Current cluster set ↕
☑ local-cluster	ManagedCluster		default
☑ sample	ManagedCluster		default

Figure 11.37 – Adding clusters to a cluster set

Finally, we can go ahead and create an ApplicationSet that uses a `Placement` object to deploy the application in all clusters that have the `env=dev` label:

```
$ oc apply -f chapter11/argocd/applicationset.yaml
applicationset.argoproj.io/cloud-api created
placement.cluster.open-cluster-management.io/cloud-api-
placement created
```

After a couple of minutes, you should see the application deployed from the application **Overview/Topology** view.

cloud-api

Actions ▼

Overview   Topology

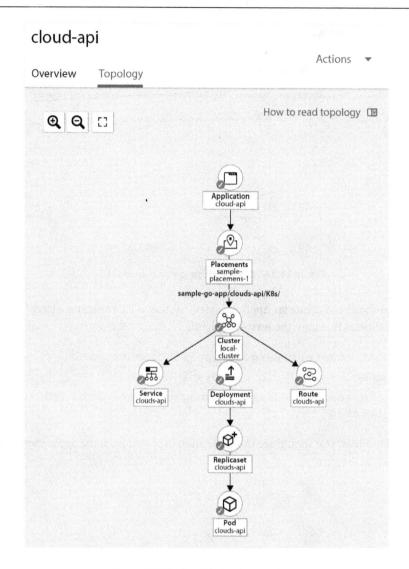

Figure 11.38 – ApplicationSet topology

The **Topology** view allows you to see your application deployed into multiple clusters from a single pane. This feature is really helpful for applications that are deployed over several clusters, as you can easily see how the application is behaving in all the clusters from a single and simple view.

This concludes our overview of the application life cycle management feature of Red Hat ACM. In this section, you have seen how ACM can help you deploy applications into multiple managed clusters by using either the Application Subscription model or OpenShift GitOps (Argo CD). Next, you are going to see how to use policies on ACM to keep your clusters compliant according to your organization's business and security needs.

Governance using Red Hat ACM

We have been discussing the challenges that large enterprises face in keeping different environments consistent a lot in this book. The ACM governance feature can play a crucial role in your strategy to maintain secure and consistent environments, no matter where they are running. The ACM governance feature allows you to define policies for a set of clusters and inform or enforce when clusters become non-compliant.

To define policies in ACM, you need to create three objects:

- **Policy**: Define the policy and remediation action that will be performed (inform or enforce).

- **PlacementBinding**: Bind a policy into a PlacementRule.

- **PlacementRule**: Rules that define which clusters the policy will be applied to.

You can see an example of a policy to check etcd encryption in all managed clusters on our GitHub. The following diagram shows what the interaction between the ACM policy objects looks like:

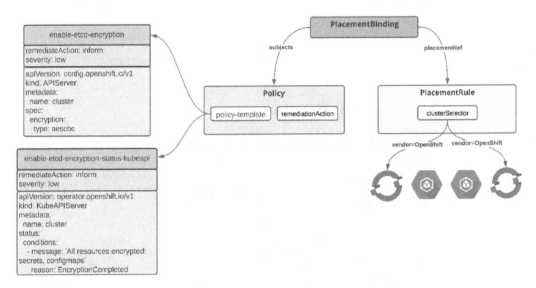

Figure 11.39 – ACM policy model

Run the following command to create the policy:

```
$ git clone https://github.com/PacktPublishing/OpenShift-Multi-
Cluster-Management-Handbook.git
$ cd OpenShift-Multi-Cluster-Management-Handbook/
$ oc apply -k chapter11/governance
namespace/acm-policies-sample created
```

```
placementrule.apps.open-cluster-management.io/placement-policy-
etcdencryption created
placementbinding.policy.open-cluster-management.io/binding-
policy-etcdencryption created
policy.policy.open-cluster-management.io/policy-etcdencryption
created
```

Now, access the **Governance** feature on the ACM web console to check the policy we just put in place.

Figure 11.40 – ACM governance console

Click on **Policies** and access **policy-etcdencryption** to see the details.

Figure 11.41 – ACM governance – violation details

In the *Further reading* section of this chapter, you will find a link to a repository that contains several reusable policies that you can use as is or as samples to create your own policies.

As you have seen, the ACM governance feature is simple to understand and use. Now think about the policies that you would like to have monitored or enforced in your clusters and start deploying your own policies!

Multi-cluster observability with Red Hat ACM

Multicluster observability is an ACM feature that is intended to be a central hub for metrics, alerting, and monitoring systems for all clusters, whether hub clusters or managed clusters.

As this tool handles a large amount of data, it is recommended to provide fast disks as its storage backend. Red Hat has tested and fully supports the solution if adopted in conjunction with Red Hat OpenShift Data Foundation.

Although Red Hat recommends doing so, the prerequisite is a storage solution that provides **object/ S3-type storage**, such as those commonly found in most cloud providers (such as Amazon S3).

Prerequisites

Since observability is a feature of an ACM operator, there aren't many prerequisites. The following are the requirements:

- Enable the observability feature on a connected Red Hat OpenShift cluster.
- Configure an object store from a storage provider. Some of the supported object storage types are as follows:

  - Red Hat Container Storage
  - AWS S3
  - Red Hat Ceph (S3-compatible API)
  - Google Cloud Storage
  - Azure Storage
  - Red Hat on IBM Cloud

> **Important Note**
>
> It is important to configure encryption when you have sensitive data persisted. The Thanos documentation has a definition of supported object stores. Check the link in the *Further reading* section at the end of this chapter.

Enabling the observability service

Since observability runs on top of ACM, its creation depends on a **Custom Resource (CR)** that will trigger the creation of the **Multicluster Observability** instance.

The following diagram demonstrates a high-level architecture of the objects involved in the observability solution. It serves as a reference for which objects are created when enabling the observability service:

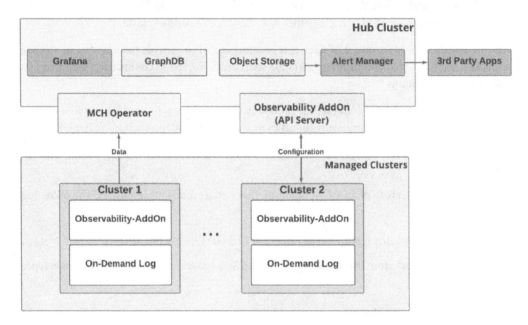

Figure 11.42 – Creating MultiClusterHub

Follow these instructions to enable multicluster observability:

1. For this example, we used an OpenShift hub cluster deployed on top of Microsoft Azure, so you must to set some variables in a terminal that will help you configure Azure dynamic storage to be used as the storage solution:

```
$ LOCATION=eastus #[1]
$ RESOURCEGROUP=aro-rg #[2]
$ CLUSTER=MyHubCluster #[3]
$ STORAGEBLOB=observsto #[4]
$ az storage account create --name $STORAGEBLOB
--resource-group $RESOURCEGROUP --location $LOCATION
--sku Standard_ZRS --kind StorageV2 #[5]
$ az ad signed-in-user show --query objectID -o
```

```
tsv | az role assignment create --role "Storage
Blob Data Contributor" --assignee @- --scope
"subscriptions/11111111-2222-a1a1-d3d3-12mn12mn12mn/
resourceGroups/$RESOURCEGROUP/providers/Microsoft.
Storage/storageAccountes/$STORAGEBLOB" #[6]
$ az storage container create --account-name $STORAGEBLOB
--name container-observ --auth-mode login #[7]
    "created": true
$ az storage account show-connection-string –name
$STORAGEBLOB
"connectionString": "DefaultEndpointsProtocol=https;End-
pointSuffix=conre.windows.net;AccountName=observsto;Ac-
countKey=sfsfoefoosdfojevntoaa/dsafojosjfsodfsafdsaf=="
#[8]
```

Let's take a look at what the highlighted numbers mean:

- #[1]: Hub cluster zone for Microsoft Azure

- #[2]: Resource group name for Microsoft Azure

- #[3]: Cluster name for the hub cluster

- #[4]: Storage account name

- #[5]: Command to create the storage on Microsoft Azure

- #[6]: Attribute storage role for OpenShift to handle storage operations for observability

- #[7]: This command creates a container inside the storage blob account, that will be used to store the data for ACM observability

- #[8]: Azure CLI instruction to get the connection string for storage

2. Create a namespace in the hub cluster for observability. We will create the namespace in the terminal; otherwise, you can also create it in the OpenShift web console UI:

```
$ oc create namespace open-cluster-management-
observability
```

3. Now, it is time to create a secret with the connection string we got in the previous step (instruction **#[8]**). The complete YAML files are available in our GitHub repository for your reference and use:

```
$DOCKER_CONFIG_JSON='oc extract secret/pull-secret -n
openshift-config --to=.' #[1]
.dockerconfigjson
$ oc create secret generic multiclusterhub-operator-pull-
secret -n open-cluster-management-observability --from-
literal=.dockerconfigjson="DOCKER_CONFIG_JSON" --type=
kubernetes.io/dockerconfigjson #[2]
(.. omitted ..)
  thanos.yaml: |
    type: AZURE
    config:
      storage_account: observsto  #[3]
      storage_account_key: sfsfoefoosdfojevntoaa/
dsafojosjfsodfsafdsaf== #[3]
      container: container-observ #[3]
      endpoint: blob.core.windows.net
      max_retries: 0
(.. omitted ..)
```

Let's take a look at what the highlighted numbers mean:

- **#[1]**: Environment variable for current Docker pull secret.
- **#[2]**: Inject the `dockerconfigjson` file into the secret for `multiclusterhub`.
- **#[3]**: Data from previous commands

Configuration file available at `https://github.com/PacktPublishing/ OpenShift-Multi-Cluster-Management-Handbook/blob/main/chapter11/ acm-observability/thanos-object-storage.yaml`.

4. Now you have the required storage to be used with observability, so it's time to create a `MulticlusterObservability` object. Go back to the **Advanced Cluster Management for Kubernetes** operator, which we installed at the beginning of this chapter, access the **MultiClusterObservability** tab, and click on the **Create MultiClusterObservability** button.

Figure 11.43 – MultiClusterObservability creation

5. Keep the CR as default and then click on **Create**.

Figure 11.44 – MultiClusterObservability

6.   Wait until the observability instance status is **Ready**.

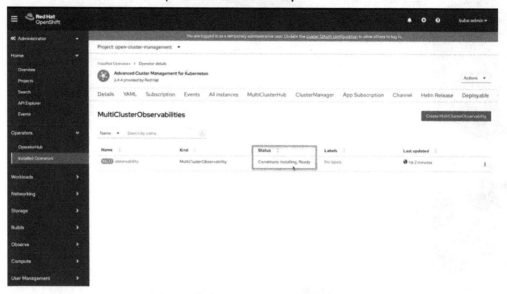

Figure 11.45 – Configuring an instance for MultiClusterObservability

7.   Now you can see, upon navigating to **Main Menu | Overview**, a route for Grafana's observability dashboard.

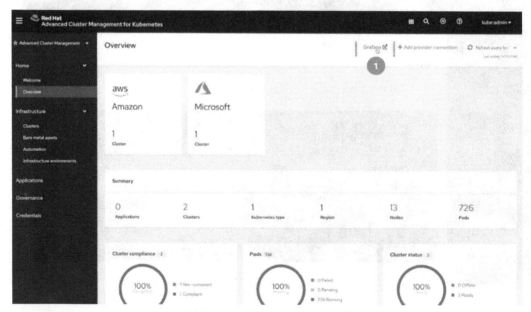

Figure 11.46 – Configuring an instance for MultiClusterObservability

#[1]: Notice that Observability option is now available in Grafana link.

8. Click on **Grafana** to see some great dashboards that aggregate metrics that come from multiple clusters.

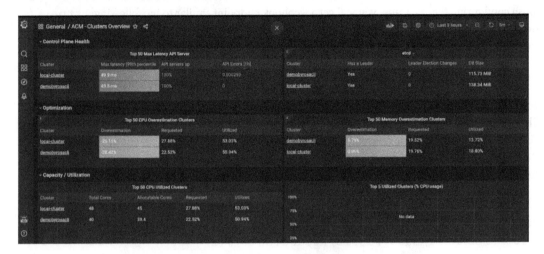

Figure 11.47 – MultiClusterObservability dashboard view sample

Now, you can count on this amazing ACM feature to help you and your organization monitor all your Kubernetes managed clusters from a central pane, independent of the infrastructure or cloud provider they are running over. In the next subsection, we will show you an option that gives you even more control over your cluster.

Configuring AlertManager to send alerts

As we have seen so far, observability can be a great ally for monitoring all your clusters from a central view, but now we will go even further and show you the *icing on the cake*, that will be one thing more to help you to manage your clusters.

As shown in *Figure 11.42*, **AlertManager** is a resource that is part of the observability architecture. We will show a sample now that you can use to enable this feature and get alerts from all managed clusters.

AlertManager is a tool that can send alerts to a set of other systems, such as email, PagerDuty, Opsgenie, WeChat, Telegram, Slack, and also your custom webhooks. For this example, we will use Slack, a short-messaging tool, as a receiver for all of our alerts.

Prerequisites

First, you will need the Slack app to set up alerts, and then point to `https://api.slack.com/messaging/webhooks` and follow the instructions to create and configure a channel. When you finish configuring the Slack app, you will get a webhook endpoint similar to the following: `https://hooks.slack.com/services/T03ECLDORAS04/B03DVP1Q91D/R4Oabcioek`. Save the webhook address in a safe place as it will be used in the next steps.

Configuring AlertManager

To configure AlertManager, you will need to create a new file named `alertmanager.yaml`. This file will have the webhook that you saved previously. The complete YAML files are available in our GitHub repository for your reference and use (github repository at `https://github.com/PacktPublishing/OpenShift-Multi-Cluster-Management-Handbook/blob/main/chapter11/acm-observability/alertmanager.yaml`):

```
global:
  slack_api_url: 'https://hooks.slack.com/services/
T03ECLDORAS04/B03DVP1Q91D/R4Oabcioek' #[1]
  resolve_timeout: 1m
route:
  receiver: 'slack-notifications'
(.. omitted ..)
  routes:
    - receiver: slack-notifications #[2]
      match:
        severity: critical|warning #[3]
receivers:
- name: 'slack-notifications'
  slack_configs:
  - channel: '#alertmanager-service' #[4]
    send_resolved: true
    icon_url: https://avatars3.githubusercontent.com/u/3380462
    title: |-
      [{{ .Status | toUpper }}{{ if eq .Status "firing" }}:{{
.Alerts.Firing | len }}{{ end }}] {{ .CommonLabels.alertname }}
for {{ .CommonLabels.job }}
(.. omitted ..)
```

In the preceding code, we have highlighted some parts with numbers. Let's take a look:

- **#[1]**: Webhook Slack API URL

- **#[2]**: Name of the receiver for alerts

- **#[3]**: Filter critical or warning alerts

- **#[4]**: Slack channel inside the workspace

The next step is to apply the new `alertmanager.yaml` file to the ACM observability namespace:

```
$ oc -n open-cluster-management-observability create secret
generic alertmanager-config --from-file=alertmanager.yaml
--dry-run=client -o=yaml |  oc -n open-cluster-management-
observability replace secret --filename=-
```

The `alertmanager.yaml` file must be in the same execution directory. Wait until the new AlertManager pods are created and you will receive new [Firing] or [Resolved] alerts on the configured channel. See an example in the following screenshot:

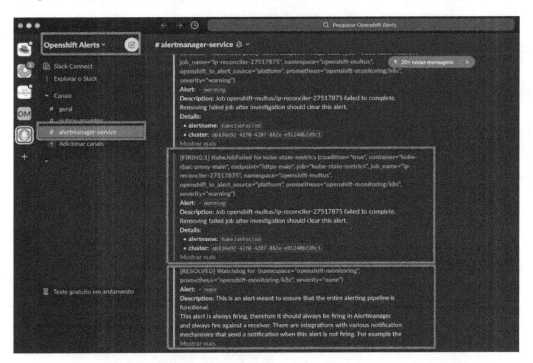

Figure 11.48 – AlertManager multicluster alerts

Here we go; we have our AlertManager set and sending alerts to a Slack channel! Therefore, in this section, you have seen the observability feature, from the installation to configuration and use. This should help you in your multi-cluster journey to monitor all your clusters, no matter which provider they are running in.

Summary

In this chapter, you have been introduced to Red Hat ACM and have seen an overview of its features and how it can help you manage several clusters. Now you understand that Red Hat ACM provides features to manage multiple clusters, keep them compliant with the policies you define for them, deploy workloads into many of them at once, and also monitor all of them from a central pane.

We also walked through the ACM installation process, provisioned a new cluster on AWS using ACM, saw how to deploy an application by using either the embedded ACM Application Subscription model or integrated with Argo CD, had a brief overview of the ACM governance feature, and, finally, enabled the observability feature to monitor multiple clusters and aggregate metrics on ACM.

In today's world, handling multiple clusters over multiple providers, either on-premises or in the cloud, is a reality in most companies; therefore, a multi-cluster management tool is a must-have. Red Hat ACM can provide you with the features you need to manage all clusters from a centralized place. We encourage you to explore and start using ACM now to reap all the benefits of this great tool.

Continue to the next chapter to learn how Red Hat Advanced Cluster Security can help you to keep your Kubernetes and OpenShift clusters secure.

Further reading

Looking for more information? Check out the following references to get more information about Red Hat ACM:

- *Red Hat Advanced Cluster Management documentation*: `https://access.redhat.com/documentation/en-us/red_hat_advanced_cluster_management_for_kubernetes/2.5`

- *Red Hat ACM supportability matrix*: `https://access.redhat.com/articles/6663461`

- *Red Hat Advanced Cluster Management observability prerequisites*: `https://access.redhat.com/documentation/en-us/red_hat_advanced_cluster_management_for_kubernetes/2.3/html-single/observability/index#prerequisites-observability`

- *Supported object stores for Thanos*: `https://thanos.io/tip/thanos/storage.md/#supported-clients`

- *Red Hat ACM policy collection:* `https://github.com/open-cluster-management/policy-collection`

- *Integrating Red Hat ACM with Ansible for cluster deployment and upgrade:* `https://access.redhat.com/documentation/en-us/red_hat_advanced_cluster_management_for_kubernetes/2.5/html/clusters/managing-your-clusters#ansible-config-cluster`

- *Integrating Red Hat ACM with Ansible for application management:* `https://access.redhat.com/documentation/en-us/red_hat_advanced_cluster_management_for_kubernetes/2.5/html/applications/managing-applications#setting-up-ansible`

- *Creating policy violation automation using Ansible:* `https://access.redhat.com/documentation/en-us/red_hat_advanced_cluster_management_for_kubernetes/2.5/html/governance/governance#configuring-governance-ansible`

Part 4 –
A Taste of Multi-Cluster
Implementation and
Security Compliance

In this part, you will take a dive into some great tools used to manage multiple clusters and to enable an organization to scale its implementation into the hybrid cloud world.

This part of the book comprises the following chapters:

12

OpenShift Multi-Cluster Security

In *Chapter 8, OpenShift Security*, we discussed some important aspects you may consider for defining and implementing security policies for your OpenShift cluster. We went through aspects such as authentication and authorization, certifications and encryption, container and network isolation, and others. If you haven't gone through that chapter yet, we encourage you to take a look now before reading this one.

Implementing security policies on OpenShift is important, but not really complicated in general – most policies' configuration is straightforward and well documented. Things become more complicated when you scale your infrastructure to several clusters though. How can you be sure that all the containers that run on top of several clusters are using secure and certified base images? Do you know how compliant all your environments are according to industry and regulatory standards such as PCI and HIPAA? To help you to monitor and maximize your OpenShift and Kubernetes clusters' security, we will introduce in this chapter Red Hat **Advanced Cluster Security** (**ACS**).

Therefore, you will find the following topics covered in this chapter:

- What is Red Hat Advanced Cluster Security?
- Red Hat Advanced Cluster Security installation
- Adding secured clusters
- Policies and violations
- Vulnerability management
- Risk profiling
- Configuration management
- Network segmentation

Let's dive into it now!

What is Red Hat Advanced Cluster Security?

Red Hat Advanced Cluster Security, also known as StackRox, is a Kubernetes-native security tool that provides the following features:

- **Policies and violations**: Defines security policies and has a report of all violation events in real time. You can define your own policies and/or use dozens of policies that come out of the box.

- **Vulnerability management**: Detects known vulnerabilities in your clusters to give you the weapons you need to remediate and prevent security issues.

- **Risk profiling**: Assesses the risk of your environment and classifies applications according to their security risk.

- **Compliance:** Reports the compliance of your clusters according to some industry-standard security profiles.

- **Configuration management**: Helps you to make sure your deployments follow the security best practices.

- **Network segmentation**: Views the network traffic between different namespaces and allows you to create Network Policies to restrict and isolate traffic.

Recent research conducted by *KuppingerCole* recognized Red Hat Advanced Cluster Security as the overall leader in the Kubernetes security segment. It is indeed a great product that is included with the **Red Hat OpenShift Plus** offering, which we will discuss in more detail in the next chapter. In the *Further reading* section of this chapter, you can find a link to *KuppingerCole's* research.

We encourage you to experiment with ACS if you have an active OpenShift Plus subscription or reach out to Red Hat's account team that covers your company for more information about it.

Red Hat Advanced Cluster Security installation

The installation process of ACS is similar to what you have seen in the last chapter with ACM: through an Operator. However, you can also install it using **Helm charts** or the **roxctl** CLI. In this book, we will use the Operator installation; if you want more information about the Helm or roxctl installation process, refer to the official documentation. You can find links to the official documentation in the *Further reading* section of this chapter.

To install ACS using the operator, proceed as follows.

Prerequisites

1. You will need access to an OpenShift cluster with cluster-admin permissions.

Operator installation

Follow the next steps to get the ACS operator ready for use:.

1. Access the OpenShift Web Console using the Administrator's perspective.
2. Navigate to the **Operators | OperatorHub** menu item.

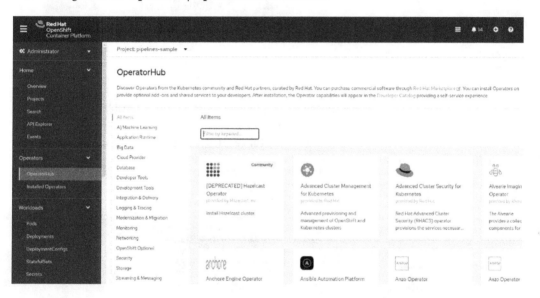

Figure 12.1 – OperatorHub

3. Search for Advanced Cluster Security for Kubernetes using the *Filter by keyword...* box.

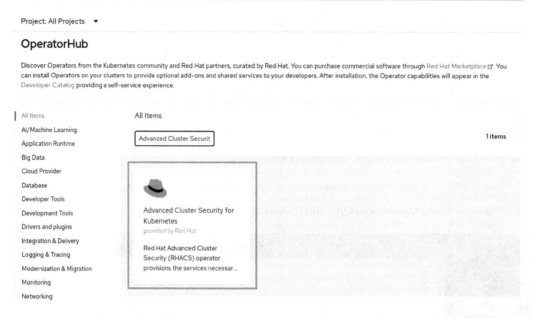

Figure 12.2 – Advanced Cluster Security for Kubernetes on OperatorHub

4. Click on the **Advanced Cluster Security for Kubernetes** tile and on the **Install** button to see the **Install Operator** screen.

Figure 12.3 – Installing Advanced Cluster Security for Kubernetes

5. Don't change the default namespace (`rhacs-operator`).

6. Select **Automatic** or **Manual** for the upgrades **Approval Strategy**. If you select **Automatic**, upgrades will be performed automatically by the **Operator Lifecycle Manager (OLM)** as soon as they are released by Red Hat, while in **Manual**, you need to approve upgrades before being applied.

7. Select the proper **Update channel**. The latest channel is recommended as it contains the latest stable and *supported* version of the operator.

8. Click the **Install** button.

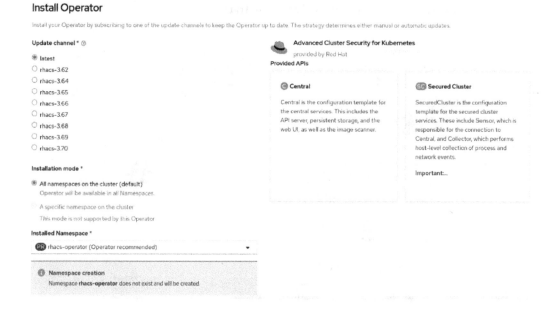

Figure 12.4 – Installing the operator

9. Wait up to 5 minutes until you see the following message:

Figure 12.5 – Operator installed

Now that we have the operator installed, the next step is to deploy the ACS Central custom resource. See next how to do it.

ACS Central installation

We can now go ahead and deploy a new ACS *Central* instance:

1. Click on the **View Operator** button or navigate to the **Operators | Installed Operators** menu and click on **Advanced Cluster Security for Kubernetes**.

2. Access the **Central** tab and click on the **Create Central** button.

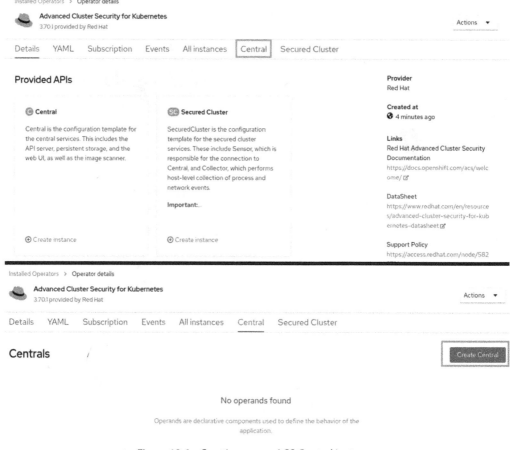

Figure 12.6 – Creating a new ACS Central instance

3. Usually, no changes are needed, so leave the default values and click on the **Create** button. Check out the link in the *Further reading* section of this chapter for product documentation for more information if you need to configure some advanced settings.

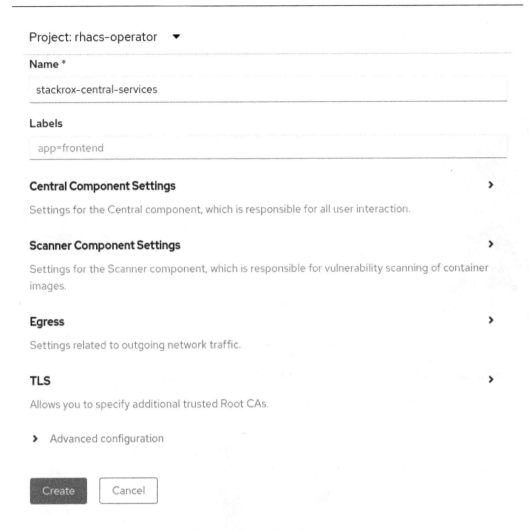

Figure 12.7 – Create ACS Central

4. Wait a few seconds until you see the **Conditions: Deployed, Initialized** status.

Figure 12.8 – ACS Central installed

5. To access the recently created ACS Central portal, you need to first get the admin credentials. To do so, navigate to **Workloads | Secrets** in the `rhacs-operator` project and click on the `central-htpasswd` secret.

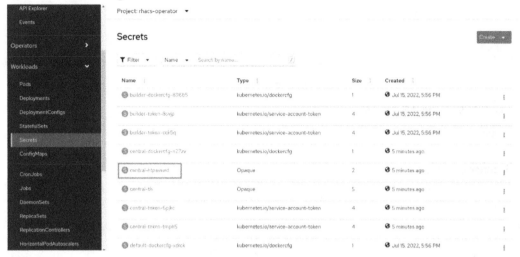

Figure 12.9 – Admin credentials secret

6. Scroll down and click on **Reveal values**. Copy the value in the **password** field.

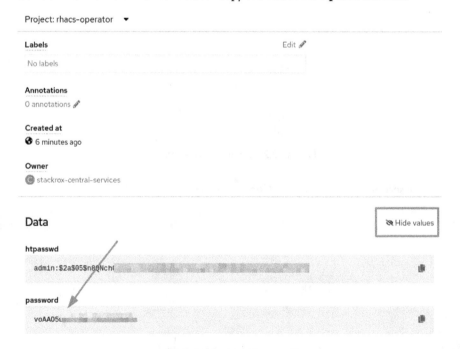

Figure 12.10 – Copy the admin password

7. Now go to **Networking** | **Routes** to get the ACS Central URL:

Figure 12.11 – ACS Central URL

8. Use the admin username and the password you copied from the secret.

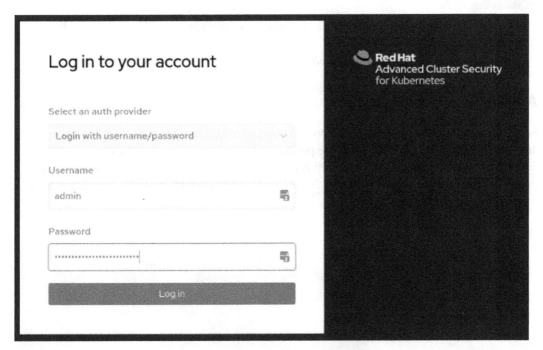

Figure 12.12 – Red Hat ACS login page

You now have Red Hat Advanced Cluster Security installed. You will notice though that we don't have any clusters under ACS management. We will add a managed cluster (also known as secured cluster) in the next steps.

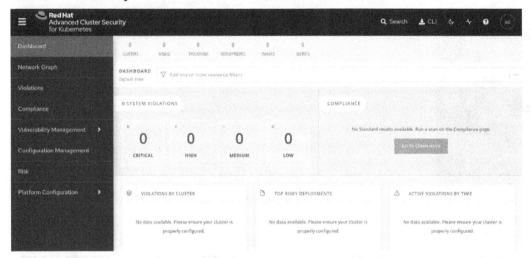

Figure 12.13 – Red Hat ACS initial page

Continue reading to see how to add secured clusters on ACS.

Adding secured clusters

Secured cluster is the term used to refer to a cluster under ACS management. ACS Central works as a control plane where you will create the policies and visualize violations, compliance, and all the features that we will walk through later in this chapter; while an ACS secured cluster is a set of ACS processes (**AdmissionControl**, **Scanner**, **Sensor**, and **Collector**) that run on managed clusters to monitor and enforce policies.

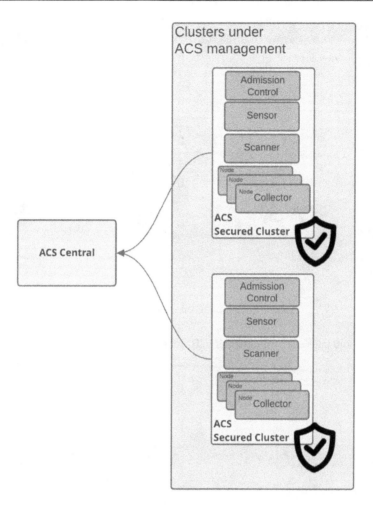

Figure 12.14 – ACS Central/secured cluster

The process of adding secured clusters on ACS Central comprises the following steps:

1. Generate an `init` bundle on ACS Central.

2. Run the `init` bundle.

3. Create a `SecuredCluster` custom resource in the ACS operator.

To perform the previous steps and add a secured cluster, run the following steps in **ACS Central**:

1. Access the **Platform Configuration | Integrations** menu:

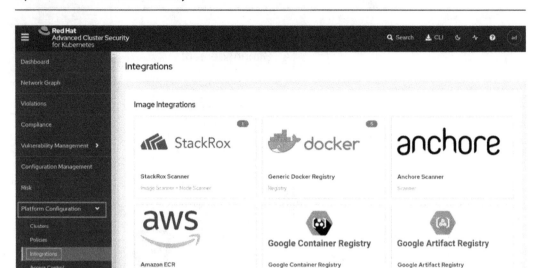

Figure 12.15 – Generating an init bundle

2. Scroll down and click on **Authentication Tokens | Cluster Init Bundle**.

Authentication Tokens

Figure 12.16 – Generating an init bundle

3. Click on **Generate bundle**, give it a name, and click on the **Download Kubernetes secrets file** button:

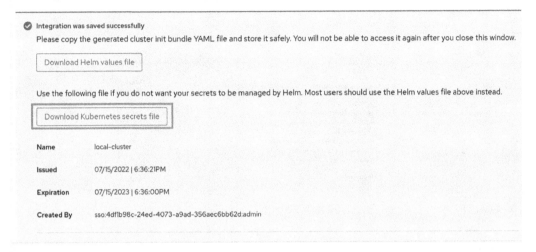

Figure 12.17 – Generating the init bundle

Now, access the OpenShift cluster that you want to add as a secured cluster (not ACS Central).

4. Using a terminal with the oc CLI, run the following command. Note that you need to run this command in the secured cluster and not on ACS Central:

```
$ oc create namespace rhacs-operator
$ oc create -f <cluster-init-secrets>.yaml -n rhacs-
operator
secret/collector-tls created
secret/sensor-tls created
secret/admission-control-tls created
```

5. Install the ACS operator in the secured cluster following the same steps we performed in the *Operator installation* section of this chapter. Note that this time, we are installing the operator in the secured cluster.

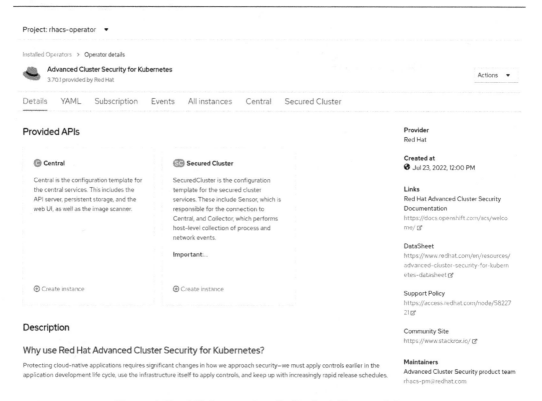

Figure 12.18 – ACS Operator installed in the ACS secured cluster

6. Still in the OpenShift secured cluster, navigate to **Operators | Installed Operators** and click on **Advanced Cluster Security for Kubernetes**. On this page, access the **Secured Cluster** tab:

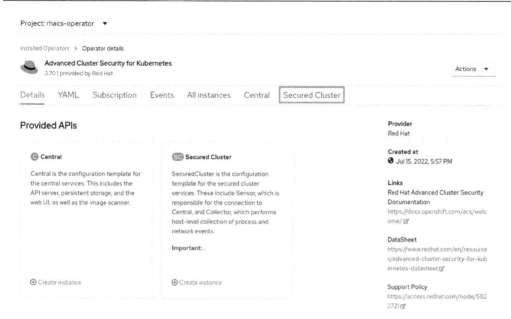

Figure 12.19 – Creating a new secured cluster

7. Click on the **Create SecuredCluster** button. On this page, give the cluster a name, add the ACS Central URL, and click on the **Create** button:

Figure 12.20 – Adding a secured cluster

> **Important Note**
>
> Always add the `port` (`:443`) in **Central Endpoint**. The secured cluster sensor and scanner may fail if you don't specify the port.

8. Wait up to 10 minutes for **ACS Collector**, **Sensor**, and **AdmissionControl** to spin up in the managed cluster. To check the status of the ACS managed cluster, access ACS Central and navigate to **Platform Configuration | Clusters**. The cluster should be marked in green as **Healthy**:

Figure 12.21 – Secured cluster health

> **Note**
>
> It is normal to see **Cluster Status** as **Degraded** during the secured cluster deployment. Wait up to 10 minutes for it to be **Healthy**. Refresh your browser to check the latest status.

We now have what we need to start using ACS: **ACS Central** monitoring one secured cluster. Continue to the next section to find out more about the ACS policies and violation features.

Policies and violations

ACS comes with dozens of security policies defined out of the box that you can just start using and also allows you to define custom security policies for your Kubernetes clusters. You can also easily check what policies are violated using the **Violations** feature.

In this section, we will see how to view and create policies and also walk through the Violations feature.

Security policies

To access the security policies, navigate to **Platform Configuration | Policies**. All out-of-the-box policies will be listed in this view:

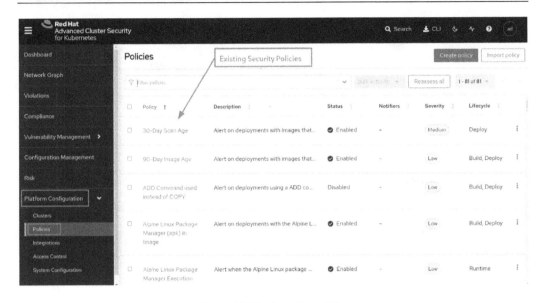

Figure 12.22 – Security policies

Let's use a simple policy to learn how a security policy works on ACS. In the **Filter policies** box, type `Policy` and hit *Enter*; then type `admin secret` and hit *Enter* again to find the `OpenShift: Advanced Cluster Security Central Admin Secret Accessed` policy:

Figure 12.23 – Latest tag policy

Now click on the link to see the **Policy details** page:

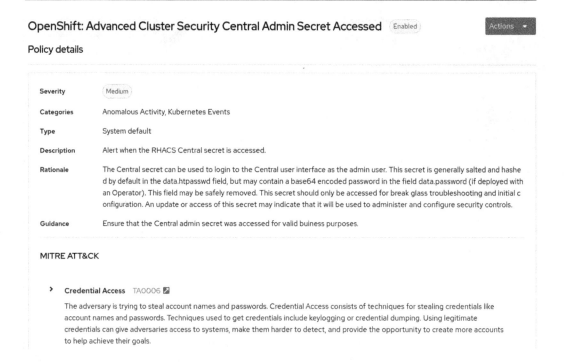

OpenShift: Advanced Cluster Security Central Admin Secret Accessed Enabled   Actions ▼

Policy details

Severity	Medium
Categories	Anomalous Activity, Kubernetes Events
Type	System default
Description	Alert when the RHACS Central secret is accessed.
Rationale	The Central secret can be used to login to the Central user interface as the admin user. This secret is generally salted and hashed by default in the data.htpasswd field, but may contain a base64 encoded password in the field data.password (if deployed with an Operator). This field may be safely removed. This secret should only be accessed for break glass troubleshooting and initial configuration. An update or access of this secret may indicate that it will be used to administer and configure security controls.
Guidance	Ensure that the Central admin secret was accessed for valid buiness purposes.

MITRE ATT&CK

> **Credential Access**   TA0006 🔗

The adversary is trying to steal account names and passwords. Credential Access consists of techniques for stealing credentials like account names and passwords. Techniques used to get credentials include keylogging or credential dumping. Using legitimate credentials can give adversaries access to systems, make them harder to detect, and provide the opportunity to create more accounts to help achieve their goals.

Figure 12.24 – Policy details

On this page, you will find policy details, such as **Description**, **Type**, and so on. Let's go ahead and click on **Actions | Edit policy** to take a look at what we can set up as part of a policy.

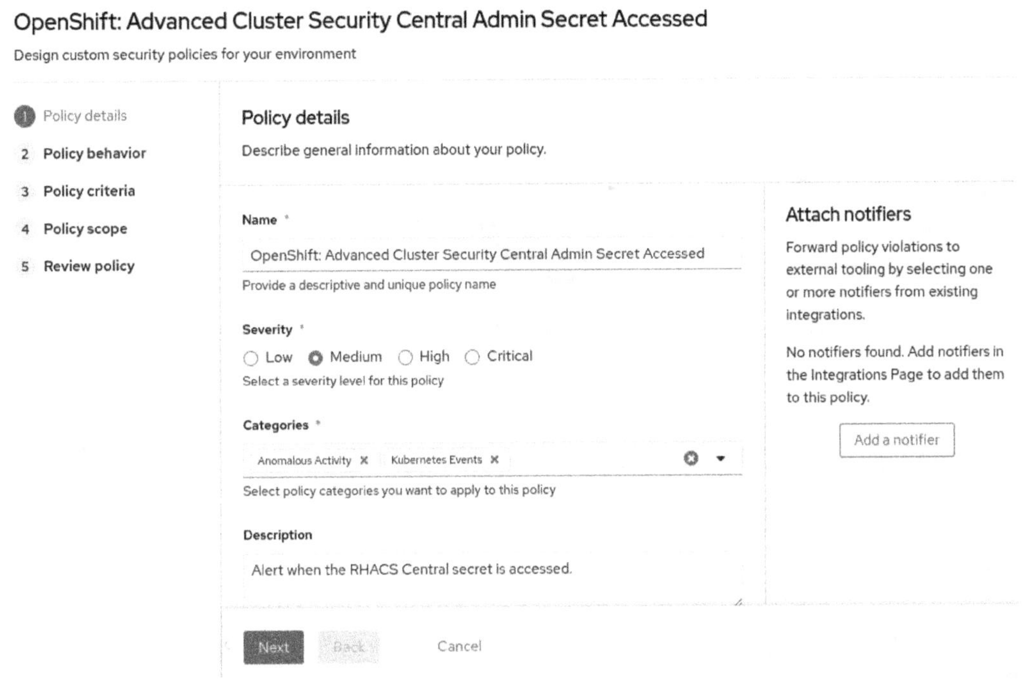

OpenShift: Advanced Cluster Security Central Admin Secret Accessed

Design custom security policies for your environment

1 Policy details

2 Policy behavior

3 Policy criteria

4 Policy scope

5 Review policy

Policy details

Describe general information about your policy.

Name *

OpenShift: Advanced Cluster Security Central Admin Secret Accessed

Provide a descriptive and unique policy name

Severity *

○ Low ● Medium ○ High ○ Critical

Select a severity level for this policy

Categories *

Anomalous Activity ✕   Kubernetes Events ✕   ⊗ ▾

Select policy categories you want to apply to this policy

Description

Alert when the RHACS Central secret is accessed.

Next   Back   Cancel

Attach notifiers

Forward policy violations to external tooling by selecting one or more notifiers from existing integrations.

No notifiers found. Add notifiers in the Integrations Page to add them to this policy.

Add a notifier

Figure 12.25 – Editing policy details

The first screen we see is **Policy details**. On this page, you can change general policy information, such as **Name**, the **Severity** level, **Categories**, and **Description**.

> **MITRE ATT&CK**
>
> MITRE ATT&CK® is a knowledge base of tactics and techniques that are used often in cyberattacks on Kubernetes. First published by Microsoft in April 2020, it is a great source of security best practices for Kubernetes. ACS allows you to classify security policies according to the MITRE ATT&CK® matrix. If you want to learn more about this framework, check the links in the *Further reading* section at the end of this chapter.

The next screen is **Policy behavior**, which defines how the policy should be applied:

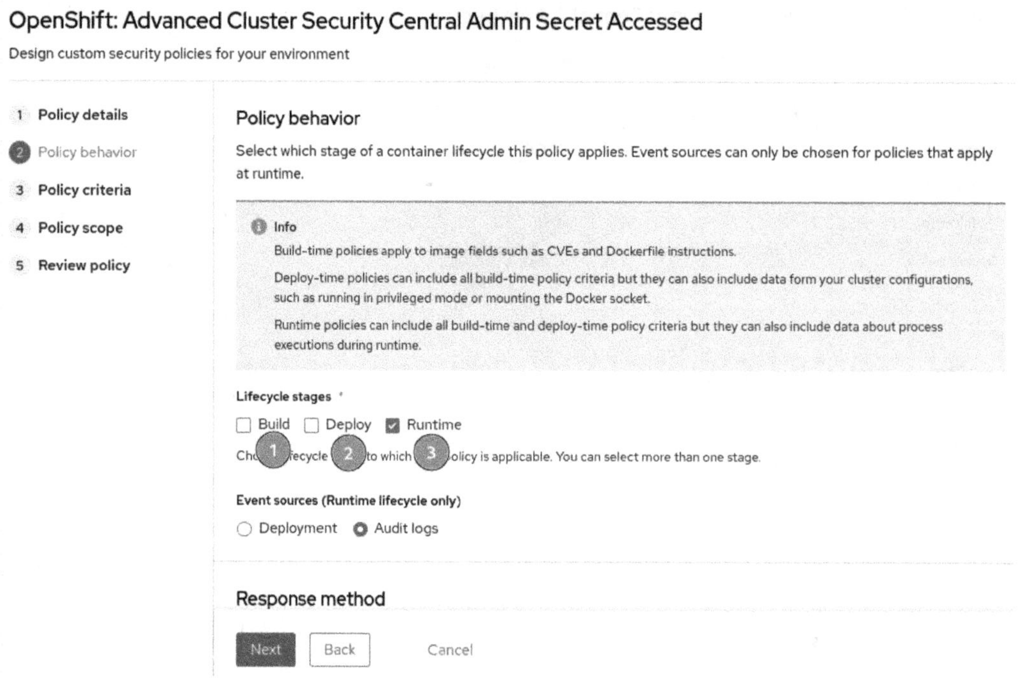

OpenShift: Advanced Cluster Security Central Admin Secret Accessed

Design custom security policies for your environment

1 Policy details

2 Policy behavior

3 Policy criteria

4 Policy scope

5 Review policy

Policy behavior

Select which stage of a container lifecycle this policy applies. Event sources can only be chosen for policies that apply at runtime.

> ⓘ **Info**
>
> Build-time policies apply to image fields such as CVEs and Dockerfile instructions.
>
> Deploy-time policies can include all build-time policy criteria but they can also include data form your cluster configurations, such as running in privileged mode or mounting the Docker socket.
>
> Runtime policies can include all build-time and deploy-time policy criteria but they can also include data about process executions during runtime.

Lifecycle stages *

☐ Build   ☐ Deploy   ☑ Runtime

Ch⬤ecycle ⬤ to which ⬤ olicy is applicable. You can select more than one stage.

Event sources (Runtime lifecycle only)

○ Deployment   ● Audit logs

Response method

[Next] [Back]   Cancel

Figure 12.26 – Policy behavior

The life cycle stages define to which stage the policy applies:

1. **Build**: Policy applied during the build of a container image. Used in general as part of a CI pipeline for static analysis of YAML Kubernetes manifests or Dockerfile instructions.

2. **Deploy**: Policies in this stage will be fired during the application deployment and will inform or even block the deployment if it violates the policy, depending on what you configured in **Response method**.

3. **Runtime**: Policy applied during runtime. For runtime policies, you can define whether **Event source** will be from **Deployment** or **Audit logs**. With the **Deployment** event source, you can inform and block application deployments that violate the policy, while **Audit logs** is used to monitor Kubernetes API calls against **Secrets** and **ConfigMaps**, to search for suspicious activities, such as sensitive passwords being read – this is exactly what is done by the policy we are using in this example.

You also can set up **Response method** for one of the following:

1. **Inform**: Only inform the violation by adding it as an item in the **Violations** feature.

2. **Inform and enforce**: Besides adding it to the violation list, it also enforces the following behavior, depending on the stage selected under Lifecycle stages:

    -   **Fails CI builds** if the **Build** stage is selected.

    -   **Blocks an application deployment** that violates the policy if the **Deploy** phase is selected.

    -   **Kills pods** that violate the policy if the **Runtime** stage is selected.

The next step describes the policy criteria that will define the policy:

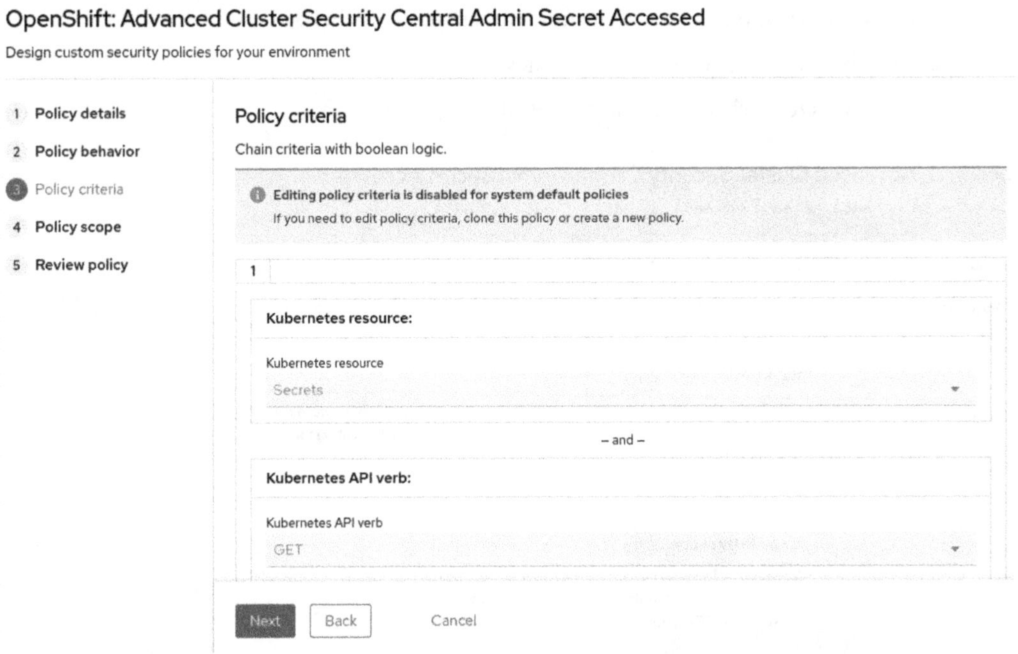

Figure 12.27 – Editing policy criteria

Policy criteria differ according to **the event source**. When the event source is **Deployment**, then you can create Boolean logic based on a large set of entities related to the image, container configuration, metadata, storage, network, container runtime processes, and Kubernetes events. We encourage you to access different policies to check the different types of policy criteria available from existing policies.

When the event source is **Audit logs**, the criteria are defined in terms of Kubernetes API events. Let's check the policy that we are using as an example to learn how audit logs-based policy criteria work. In our example, the following criteria are used:

-   Kubernetes resource: **Secrets**
-   Kubernetes API verb: **GET**, **PATCH**, or **UPDATE**

- **Kubernetes resource name:** `central-htpasswd`
- **Kubernetes user name is NOT:** `system:serviceaccount:openshift-authentication-operator:rhacs-operator-controller-manager`

This means that a violation will be raised when the `central-htpasswd` secret is either **read** (`GET`) or **changed** (`PATCH` or `UPDATE`) by *any user other than the* `rhacs-operator-controller-manager service account`.

You can also set the policy scope if you want:

1. Restrict the policy to only specific clusters, namespaces, or labels.
2. Alternatively, exclude entities from the policy.
3. Exclude images to be checked during the **Build** stage:

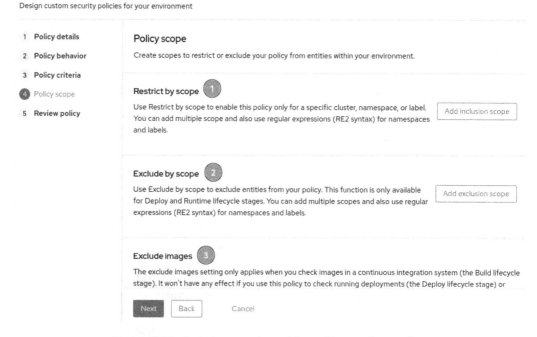

Figure 12.28 – Exclude or restrict entities and images from policy

This concludes the configurations you will find in a policy. It is not that difficult, right? We encourage you to create some custom policies to practice and learn from them.

Violations

The **Violations** feature lists all security policies that have been violated in the clusters monitored by ACS:

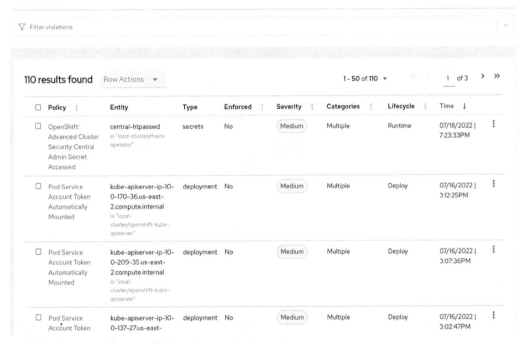

Figure 12.29 – Violations list

Remember that we read `central-htpasswd` in *step 15* of the *ACS Central installation* section to get the ACS Central admin password. That raised a violation due to the policy we used as an example previously (GET *API of* `central-htpasswd` *secret*). Click on some of the violations you see on this page to explore the feature and learn about the events and data that are captured and shown by the ACS **Violations** feature:

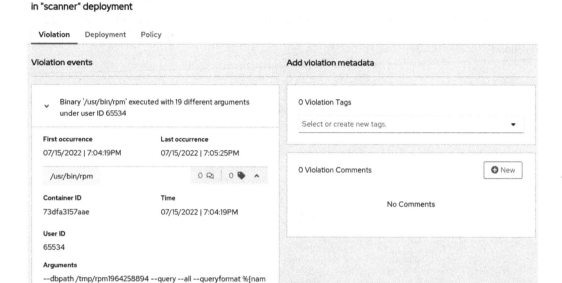

Red Hat Package Manager Execution
in "scanner" deployment

Violation   Deployment   Policy

Violation events

Binary '/usr/bin/rpm' executed with 19 different arguments under user ID 65534

First occurrence   Last occurrence
07/15/2022 | 7:04:19PM   07/15/2022 | 7:05:25PM

/usr/bin/rpm

Container ID   Time
73dfa3157aae   07/15/2022 | 7:04:19PM

User ID
65534

Arguments
--dbpath /tmp/rpm1964258894 --query --all --queryformat %{name}\n%{evr}\n%{ARCH}\n%{RPMTAG_MODULARITYLABEL}\n[%{FILENAMES}\n].\n

Add violation metadata

0 Violation Tags

Select or create new tags.

0 Violation Comments   ⊕ New

No Comments

Figure 12.30 – Example of a violation

In this section, we have learned how a security policy is defined on ACM and how to easily see a list of the violations that occur in your clusters. Continue reading to learn more about the **Vulnerability Management** feature of ACS and how you can use it to identify, prioritize and remediate vulnerabilities.

Vulnerability management

There is a general consensus that any system has vulnerabilities; some of them are known and some are not identified yet. Vulnerability management is the process of identifying and managing known vulnerabilities, which means having plans in place to remediate or mitigate the impact of the vulnerabilities. Navigate to **Vulnerability Management** | **Dashboard** to see what this feature looks like:

Figure 12.31 – Vulnerability management

Through this feature, you can walk through all the vulnerabilities detected by ACS and decide what actions to take:

- Remediate the vulnerability either by removing the vulnerable software package from the application or updating it with a more recent version in which the vulnerability is already fixed.

- Accept the risk.

- Mark it as a false positive.

Vulnerabilities are detected and grouped in terms of the following:

- **Components**: Software packages used by containers. This group helps you to detect the software packages that contain more vulnerabilities and where they are used, so you can upgrade the applications accordingly to remediate them.

- **Image**: Group the vulnerabilities by images. Similarly, you can view what images are more vulnerable, check whether there is a fix for the security issues, and plan accordingly.

- **Nodes**: Group the vulnerabilities by nodes.

- **Deployment**: See the vulnerabilities by deployment. Easier to check vulnerabilities for specific applications.

- **Namespace**: Vulnerabilities by namespace.

- **Cluster**: All vulnerabilities by clusters.

These groups are accessed from the buttons at the top of the **Vulnerability Management** dashboard; click on them to explore the different ways you can see and filter the vulnerabilities:

Figure 12.32 – Group by entities

We are going to deploy a sample application now to check the **Vulnerability Management** feature in action. To do so, run the following commands in the OpenShift secured cluster:

```
$ oc new-project acs-test
$ oc run samba --labels=app=rce \
    --image=vulnerables/cve-2017-7494 -n acs-test
$ oc run shell --labels=app=shellshock,team=test-team \
    --image=vulnerables/cve-2014-6271 -n acs-test
```

Now, access again the **Vulnerability Management** dashboard. You may notice some interesting things in the dashboard now:

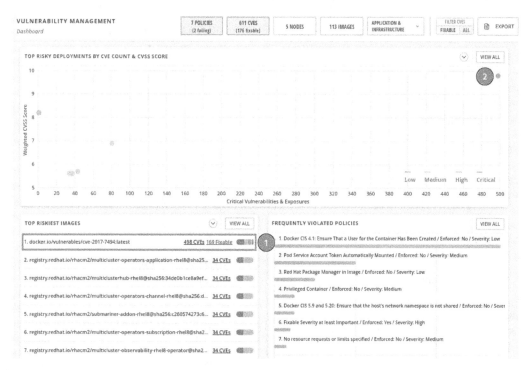

Figure 12.33 – Top riskiest images and deployments

In the previous screenshot, you can see the following:

- The container image from the application we just deployed is listed as the top riskiest image.

- In the graph, the application deployment is shown as the most critical in terms of CVE count and CVSS score.

Click on **APPLICATION & INFRASTRUCTURE | Namespaces** to view the vulnerabilities grouped by namespaces:

Figure 12.34 – APPLICATION & INFRASTRUCTURE menu

You will notice now our `acs-test` namespace is listed with more than 400 CVEs, with more than 150 fixable:

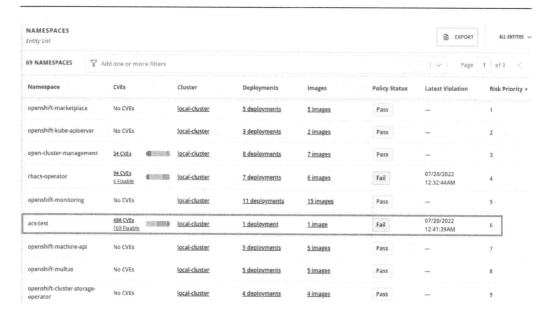

Figure 12.35 – Vulnerabilities by namespace

Click on the `acs-test` namespace to drill down and see the details.

Figure 12.36 – Namespace Summary view

Explore the **Related entities** menu on the right side of the screen to check that you can easily find the CVEs for the `samba` deployment, policies that pass or fail for the namespace, a list of images and their vulnerabilities, and all components (software packages) that have CVEs, and, finally, a list of all CVEs detected in that namespace. You can learn by exploring the namespace that it contains two applications (deployments), in which one of them has 498 known CVEs, of which 169 are fixable:

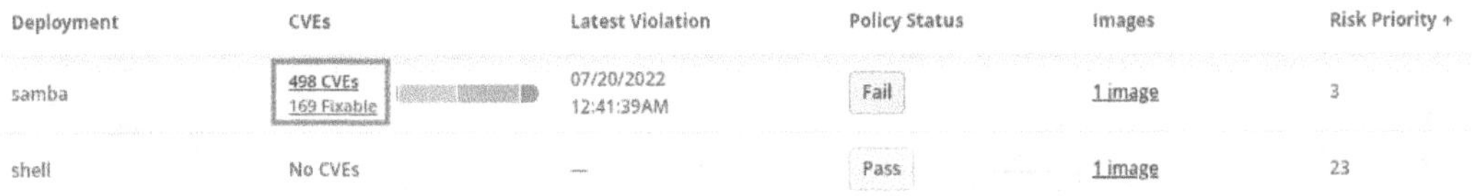
Deployment
CVEs
Latest Violation
Policy Status
Images
Risk Priority
samba
498 CVEs
169 Fixable
07/20/2022
12:41:39AM
Fail
1 image
3
shell
No CVEs
—
Pass
1 image
23

Figure 12.37 – CVEs detected

By drilling down into the deployment and components, you can check which package version is in use and in which version the CVE was fixed:

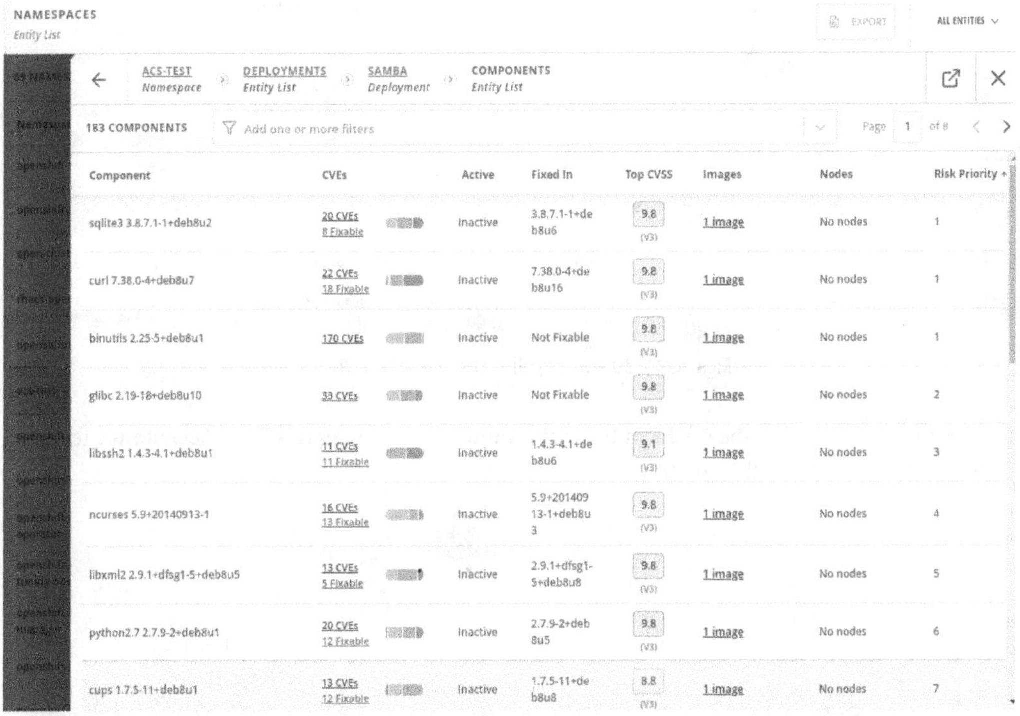
NAMESPACES
Entity List
EXPORT
ALL ENTITIES
ACS-TEST
Namespace
DEPLOYMENTS
Entity List
SAMBA
Deployment
COMPONENTS
Entity List
183 COMPONENTS
Add one or more filters
Page 1 of 8
Component
CVEs
Active
Fixed In
Top CVSS
Images
Nodes
Risk Priority
sqlite3 3.8.7.1-1+deb8u2
20 CVEs
8 Fixable
Inactive
3.8.7.1-1+deb8u6
9.8 (V3)
1 image
No nodes
1
curl 7.38.0-4+deb8u7
22 CVEs
18 Fixable
Inactive
7.38.0-4+deb8u16
9.8 (V3)
1 image
No nodes
1
binutils 2.25-5+deb8u1
170 CVEs
Inactive
Not Fixable
9.8 (V3)
1 image
No nodes
1
glibc 2.19-18+deb8u10
33 CVEs
Inactive
Not Fixable
9.8 (V3)
1 image
No nodes
2
libssh2 1.4.3-4.1+deb8u1
11 CVEs
11 Fixable
Inactive
1.4.3-4.1+deb8u6
9.1 (V3)
1 image
No nodes
3
ncurses 5.9+20140913-1
16 CVEs
13 Fixable
Inactive
5.9+20140913-1+deb8u3
9.8 (V3)
1 image
No nodes
4
libxml2 2.9.1+dfsg1-5+deb8u5
13 CVEs
5 Fixable
Inactive
2.9.1+dfsg1-5+deb8u8
9.8 (V3)
1 image
No nodes
5
python2.7 2.7.9-2+deb8u1
20 CVEs
12 Fixable
Inactive
2.7.9-2+deb8u5
9.8 (V3)
1 image
No nodes
6
cups 1.7.5-11+deb8u1
13 CVEs
12 Fixable
Inactive
1.7.5-11+deb8u8
8.8 (V3)
1 image
No nodes
7

Figure 12.38 – Components and CVEs

That gives a lot of helpful information for you to identify vulnerable packages and images and remediate them. You can alternatively also decide to accept the risk (defer) or mark it as a false positive; to do so, access the image, scroll down and select the CVEs you want to defer, and click on **Defer CVE** or **Mark false positive**:

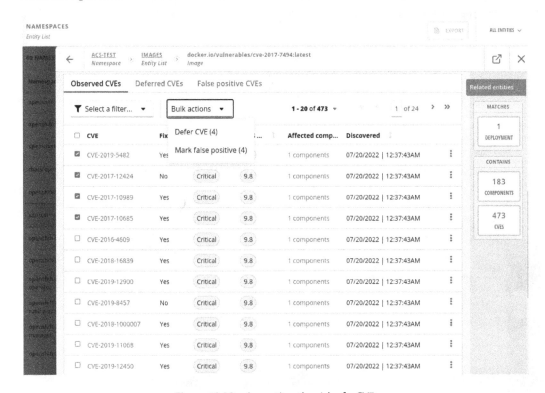

Figure 12.39 – Accepting the risk of a CVE

You can review and approve the CVEs in the **Vulnerability Management | Risk Acceptance** feature and also list the approved deferrals and false positives:

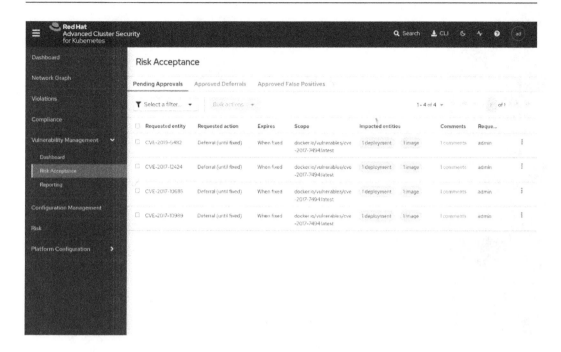

Figure 12.40 – Risk Acceptance

The **Vulnerability Management** feature also includes a report feature, in which you can set up a security report and send it weekly or monthly to a distribution list. It is helpful to report vulnerabilities across the organization frequently. We will not cover this feature in this book, but you can find a link in the *Further reading* section of this chapter that will help you to configure it.

Risk profiling

The **Risk** view is a feature that classifies all running deployments in terms of security risks. Navigate to the **Risk** menu to access the feature and learn from it:

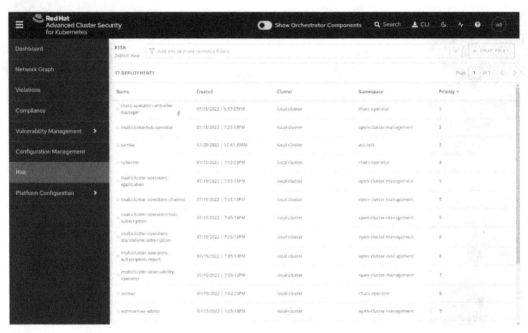

Figure 12.41 – Risk feature

Click on the samba deployment and explore **RISK INDICATORS**, **DEPLOYMENT DETAILS**, and **PROCESS DISCOVERY**:

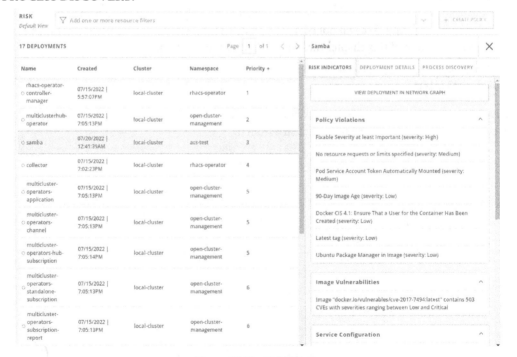

Figure 12.42 – Risk details

We will not walk through each option there but will highlight the **PROCESS DISCOVERY** tab, which provides some interesting insights, as you can see next:

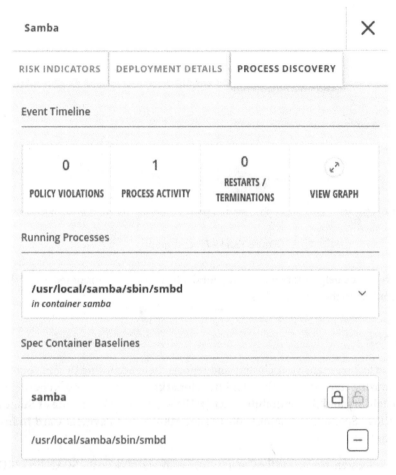

Figure 12.43 – Process Discovery

In this tab, you can see all the processes that are running in the container including details and also a graph that shows the process activities over time. Click on the **VIEW GRAPH** link to see it:

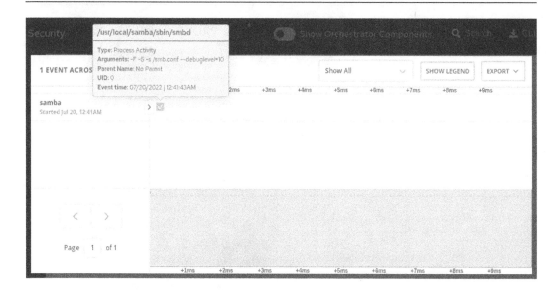

Figure 12.44 – Process graph

This **Risk** feature can be helpful for you to sort your deployments in terms of security risks and then take action according to the prioritized list.

Compliance

The **Compliance** feature scans your clusters and reports them according to some out-of-the-box compliance industry standards, such as CIS Benchmarks for Docker and Kubernetes, **Health Insurance Portability and Accountability Act (HIPAA)**, **National Institute of Standards and Technology (NIST)** Special Publications 800-190 and 800-53, and **Payment Card Industry Data Security Standard (PCI DSS)**.

To run the compliance scan, navigate to the **Compliance** feature and click on the **SCAN ENVIRONMENT** button:

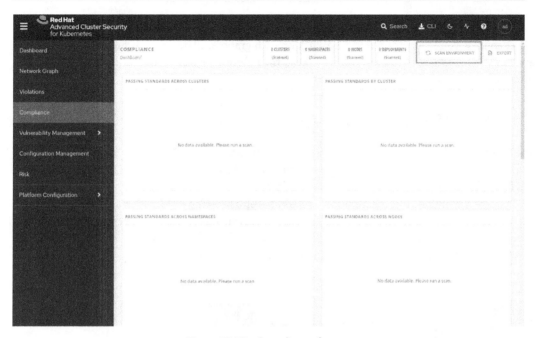

Figure 12.45 – Compliance feature

After some seconds, you will see the compliance report, as follows:

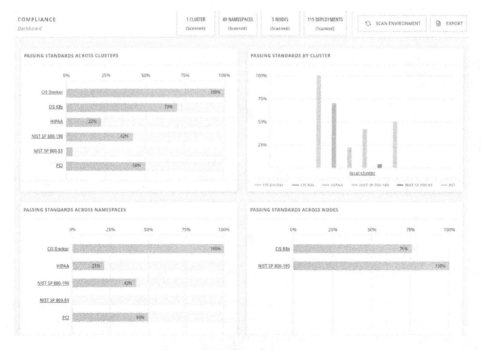

Figure 12.46 – Compliance report

We will not dive into each of these industry standards, as they are very specific to different industries. We encourage you to explore the feature, click on each graph, drill down, and check which controls are compliant and not compliant and why. Read the reference we left in the *Further reading* if you want to see more about this compliance feature.

Configuration Management

The **Configuration Management** feature is a different way to look at policies that are violated, correlating with the various objects that make up the configuration and use of the clusters. Using this feature, you can list all failing policies for all clusters and drill down to inspect all namespaces, deployments, and so on. You may be thinking that the same information can also be found in **Violations** and **Vulnerability Management**, and that is true! The same information is also there; however, here, you will find it grouped by clusters' entities and also displaying summary data of each entity, which will help you to correlate the different entities and learn about the connections among them.

To access this feature, navigate to the **Configuration Management** menu:

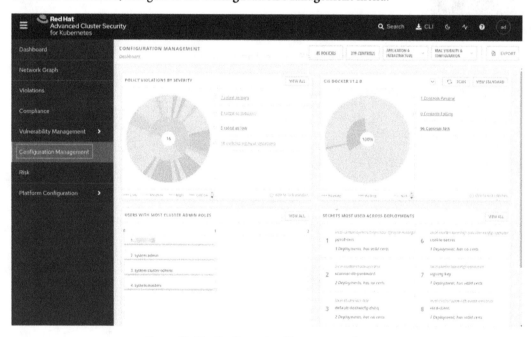

Figure 12.47 – Configuration Management feature

You will first see a dashboard that summarizes the following information:

- **POLICY VIOLATIONS BY SEVERITY**: Group the policy violations by severity (critical, high, medium, and low) and display them in a pie chart. You can drill down and inspect the policies by clicking on the link on the right side of the pie chart:

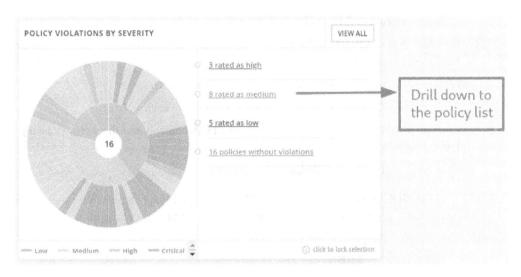

Figure 12.48 – POLICY VIOLATIONS BY SEVERITY pie chart

- **CIS DOCKER/Kubernetes**: Another pie chart grouping entities by CIS Docker or CIS Kubernetes controls:

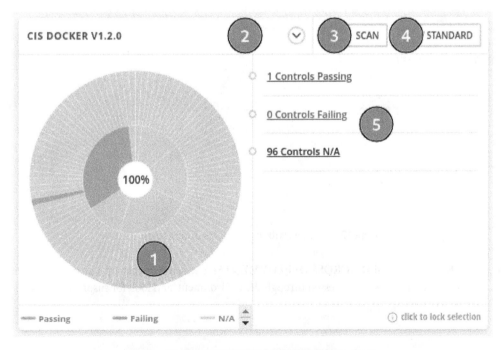

Figure 12.49 – CIS DOCKER/KUBERNETES pie chart

See next the description of this pie chart:

1.   Pie chart that summarizes the controls compliance considering the CIS Docker or Kubernetes standard.

2.   Click here to switch between CIS Docker and CIS Kubernetes.

3.   Click on the **SCAN** button to perform a new scan in the environment.

4.   Click on **VIEW STANDARD** to get a list of all CIS controls and the status (**Passing, Failing**, or **N/A**).

5.   Drill down to the passed, failed, or N/A control list.

- **USERS WITH MOST CLUSTER ADMIN ROLES**: This is self-explanatory. You can use this list to review users' permissions and make sure they have the proper permissions:

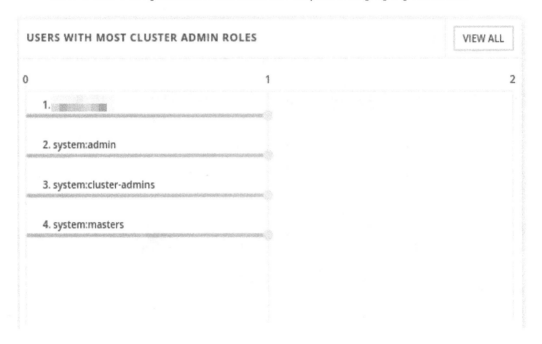

Figure 12.50 – Users with most cluster admin roles

- **SECRETS MOST USED ACROSS DEPLOYMENTS**: Also self-explanatory. Use this list to find sensitive data and how it is accessed through the environment, and look for suspicious activities:

SECRETS MOST USED ACROSS DEPLOYMENTS   VIEW ALL

local-cluster/openshift-operator-lifecycle-manage   *local-cluster/openshift-machine-config-operator*

1   pprof-cert   6   cookie-secret

3 Deployments, has valid certs   *1 Deployment, has no certs*

local-cluster/rhacs-operator   *local-cluster/openshift-service-ca*

2   scanner-db-password   7   signing-key

2 Deployments, has no certs   *1 Deployment, has valid certs*

local-cluster/acs-test   *local-cluster/openshift-oauth-apiserver*

3   default-dockercfg-zlxhq   8   etcd-client

2 Deployments, has no certs   *1 Deployment, has valid certs*

local-cluster/openshift-machine-config-operator   *local-cluster/openshift-apiserver*

4   machine-config-server-tls   9   etcd-client

1 Deployment, has valid certs   *1 Deployment, has valid certs*

local-cluster/openshift-cluster-csi-drivers   *local-cluster/openshift-authentication*

5   ebs-cloud-credentials   10 v4-0-config-system-ocp-branding-templa

1 Deployment, has no certs   *1 Deployment, has no certs*

Figure 12.51 – Secrets most used across deployments

You can also use the bar at the top of this page to inspect the policies and controls by clusters, namespaces, nodes, deployments, images, secrets, users and groups, service accounts, and roles:

Figure 12.52 – View policies and controls by entities

Click on the menu options and explore the different lists you get with each of them. Notice also that you can drill down from clusters to deployments, images, and so on to navigate back and forth to analyze the entities:

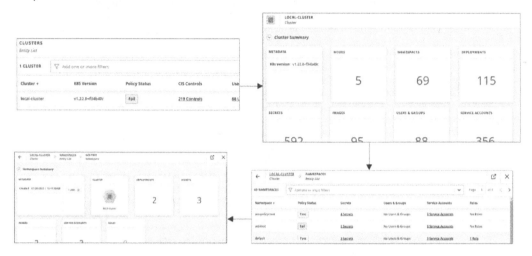

Figure 12.53 – Drill down from cluster to namespace

This concludes an overview of the **Configuration Management** feature. Continue reading to see a great feature ACS brings to help you to inspect your clusters' network communication.

Network segmentation

An important security aspect in any Kubernetes cluster is how Pods communicate between each other and also ingress and egress communication. Currently, there isn't any graphical view on Kubernetes to check how the network communications are performed in real time, and neither allowed nor blocked flows. To help with that, ACS brings the **Network Graph** feature, which allows you to view the active communications in real time and also define and apply NPs to allow or block network traffic. Click on the **Network Graph** menu to access the feature:

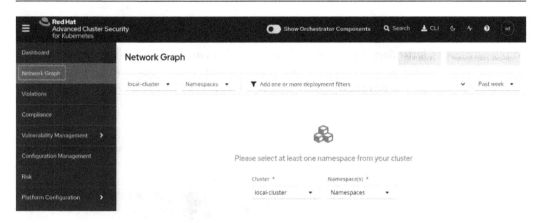

Figure 12.54 – Network Graph feature

Select the `rhacs-operator` namespace to view what the network graph looks like:

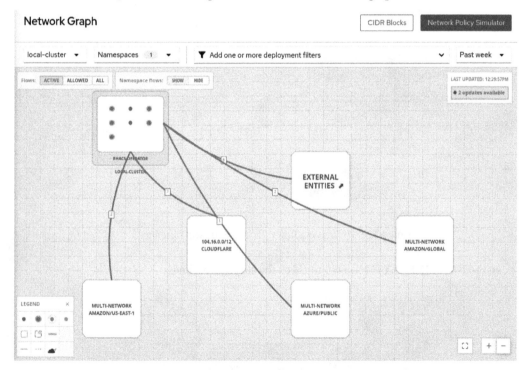

Figure 12.55 – Network graph for the rhacs-operator namespace

You can change the view to see only active connections, allowed connections, or all types of connection flows.

Network flows

Red Hat Advanced Cluster Security can learn the network flows used by the applications and apply a baseline of all network flows. Any network flows detected that are different from the baseline are marked as anomalous for your review. When viewing the network flows and baseline, access any deployment, and you will see the anomalous flows marked in red:

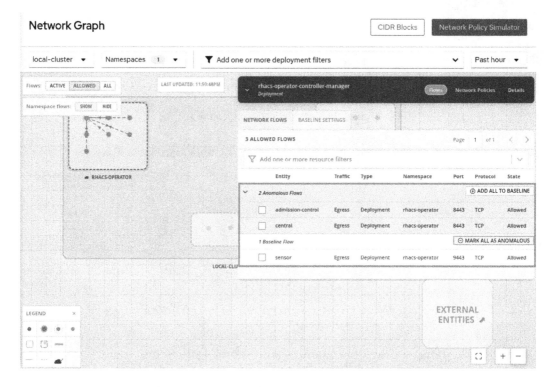

Figure 12.56 – Network flows – anomalous and baseline

Click on the **BASELINE SETTINGS** tab to check the current baseline:

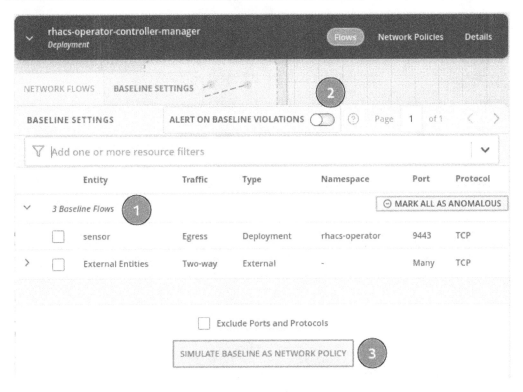

Figure 12.57 – Configure baseline

Through the **BASELINE SETTINGS** tab, you can do the following:

1. View the baseline of network flows.

2. Configure sending an alert when ACS detects anomalous flows.

3. Simulate the impact in the environment of having the baseline as NPs.

See next how to also use the **Network Graph** feature to generate a list of **NP**s (Network Policies) to allow only the communications that are required by the applications.

Network Policy Simulator

Do you know whether the NP configurations of your clusters allow only required communications and nothing more? Clusters with a permissive set of NPs are very common and the ACS Network Policy Simulator can help to avoid that. ACS monitors the network traffic between all Pods and namespaces in the clusters to create a matrix of firewall rules. You can use ACS to generate a set of NPs based on what it learned from the environment that will allow only the communications needed. To use this feature, click on the **Network Policy Simulator** button then click on **Generate and simulate network policies**:

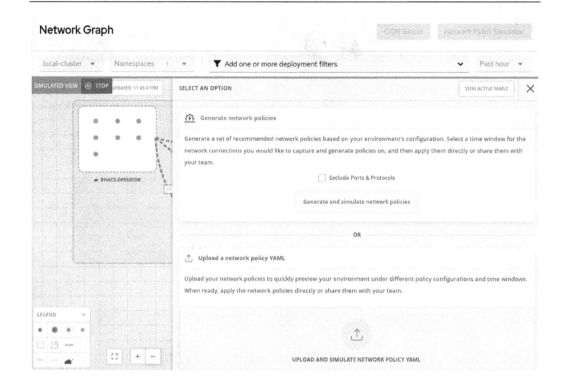

Figure 12.58 – Generate and simulate network policies

You will see an extensive list of NPs that will allow only the communications that ACS learned from the environment that is in use. You can apply the NPs in the environment or share them by email. We highly recommend you review and test the NPs in a development or test environment before applying them in production.

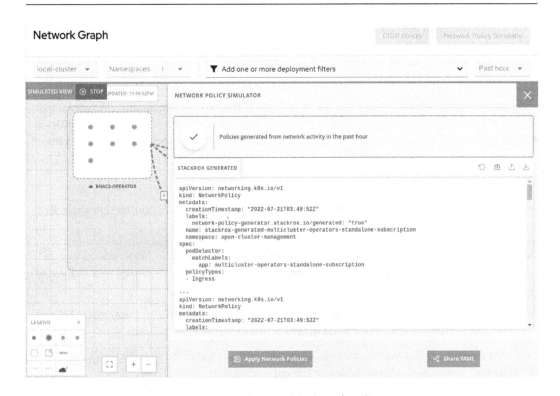

Figure 12.59 – Apply or share NPs

Helpful, right? Now explore the **Network Graph** feature a bit more and envision how this feature can help you and your organization to have more visibility into the network traffic and also allow only the communications that are really required by your applications.

Summary

We covered a lot of content in this chapter about Red Hat Advanced Cluster Security. In this chapter, we have seen an overview of ACS capabilities to help you to learn how ACS can help to make sure your clusters are secure and vulnerabilities are known, and put an action plan in place.

We learned how to use and define security policies and list all policy violations using the **Violations** feature. We also saw that the **Vulnerability Management** feature is very helpful to list all known vulnerabilities, review them, and take proper action: remediate (fix vulnerable packages), accept the risk, or mark them as false positives.

We also learned that the **Risk** profiling feature helps you to assess the risk of application deployments and prioritize the remediation and actions that need to be taken to enhance the security. **Compliance** reports the clusters, namespaces, and deployments in terms of industry standards, such as CIS Docker, HIPAA, NIST, PCI, and so on.

Finally, we saw the list of policies and controls aggregated by a cluster's entities in the **Configuration Management** feature, helping to correlate the different entities into the clusters. **Network Graph**, in turn, gave us a nice view of the network flows in real time with some useful added features to help generate and simulate NPs and make sure only needed communications are allowed and nothing more.

We hope this chapter helped you to understand Red Hat Advanced Cluster Security. We encourage you to move on to the next chapter to see how ACM, ACS, and other pieces will build together a complete and comprehensive platform for multi-cloud or hybrid cloud: **OpenShift Platform Plus**.

Further reading

Looking for more information? Check the following references to get more information about Red Hat Advanced Cluster Security:

- *KuppingerCole Report Leadership Compass: Container security*: https://www.redhat.com/en/resources/kuppingercole-container-security-report-analyst-material

- *ACM installation using the Operator*: https://docs.openshift.com/acs/3.70/installing/install-ocp-operator.html

- *ACM installation using Helm*: https://docs.openshift.com/acs/3.70/installing/installing_helm/install-helm-quick.html

- *ACM installation using the roxctl CLI*: https://docs.openshift.com/acs/3.70/installing/install-quick-roxctl.html

- *Threat matrix for Kubernetes* – Microsoft article about MITRE ATT&CK®: https://www.microsoft.com/security/blog/2020/04/02/attack-matrix-kubernetes/

- *Protecting Kubernetes Against MITRE ATT&CK: Initial Access* – First of a series of articles about the MITRE ATT&CK® framework: https://cloud.redhat.com/blog/protecting-kubernetes-against-mitre-attck-initial-access

- *MITRE ATT&CK® knowledge base*: https://attack.mitre.org/

- *Configuring the vulnerability reports*: https://docs.openshift.com/acs/3.70/operating/manage-vulnerabilities.html#vulnerability-management-reporting_acs-operating-manage-vulnerabilities

- *Managing compliance*: https://docs.openshift.com/acs/3.70/operating/manage-compliance.html

13
OpenShift Plus – a Multi-Cluster Enterprise Ready Solution

In this book, we discussed in *Chapter 1, Hybrid Cloud Journey and Strategies*, the main challenges related to public cloud usage, such as keeping cloud costs under control and having consistent, secure, and compliant platforms, no matter which cloud provider is being used. We also introduced some of the tools that are helpful to work with multi-clusters and help to address those challenges, from a technology perspective.

In *Chapter 9, OpenShift Pipelines – Tekton*, and *Chapter 10, OpenShift GitOps – ArgoCD*, we covered **CI/CD pipelines** and **GitOps** using OpenShift, including deployment into multiple clusters.

In *Chapter 11, OpenShift Multi-Cluster GitOps and Management*, we saw how to use **Red Hat Advanced Cluster Management** to manage and observe several clusters from a single pane, deploy workloads into many clusters at once, and also use policies to ensure all your clusters are compliant.

Finally, in the previous chapter – *Chapter 12, OpenShift Multi-Cluster Security* – we introduced the **Red Hat Advanced Cluster Security** tool, which helps to enhance the security of all your clusters, by implementing security policies, managing the known vulnerabilities, assessing the security risks and compliance, and managing the network traffic.

Now, we have two different focuses in this chapter, the first is to introduce the last important piece for the hybrid cloud strategy: **Red Hat Quay**, which will store all container images in a central image registry and makes your CI/CD process a bit easier and more robust; and the second is to discuss the **Red Hat OpenShift Plus** offering, which bundles all these pieces together to provide you a comprehensive and competitive solution.

Therefore, you will find the following in this chapter:

- Introducing Red Hat Quay
- Deploying Red Hat using the Quay Operator
- Using Red Hat Quay
- What is OpenShift Plus?
- OpenShift Plus: A practical use-case

Let's start now then!

Introducing Red Hat Quay

Red Hat Quay is an enterprise container registry platform that runs on Red Hat Enterprise Linux or OpenShift, on-premise or in the cloud. Red Hat Quay provides great features for an image registry, such as the following:

- **Image vulnerability scan**: Scan the images to find known vulnerabilities just after they are pushed to the registry using the Clair project. See more information about Clair in the *Further reading* section of this chapter.

- **Geo-replication**: Sync the image registry contents between two or more Quay instances, allowing multiple and geographically distributed Quay deployments to look like a single registry. This is especially helpful for environments spread over far distances and with high latency. Quay handles asynchronous data synchronization between the different instances to make images available transparently for the end user.

- **Repository mirroring**: This synchronizes one or more repositories between two Quay instances.

- **Access control**: Granular permission control allows you to define who can read, write, and administer groups and repositories. You can also leverage robot accounts to allow automated access to organizations and repositories using the Quay API.

Red Hat Quay is available as a managed offering on the quay.io portal or self-managed, where you deploy and maintain it.

Deploying Red Hat Quay using the Quay Operator

There are different ways to deploy Quay, as you can see in the product's official documentation link that you'll find in the *Further reading* section of this chapter. For didactical reasons, we decided to deploy it using the Quay Operator with fully-managed components – a comfortable way to start using Quay.

Some prerequisites are necessary to install Quay as an enterprise container registry.

Prerequisites

The prerequisites to install Quay using the Operator are as follows:

- An OpenShift cluster with a privileged account to deploy and set up Quay.
- Object storage: The supported object storage is Red Hat OpenShift Data Foundation, AWS S3, Google Cloud Storage, Azure Storage, Ceph/RadosGW Storage/Hitachi HCP storage, Swift storage, and NooBaa.
- Cluster capacity to host the following services:
  - PostgreSQL or MySQL database. PostgreSQL is preferred due to enhanced features for Clair security scanning.
  - A proxy server.
  - A key-value database. Redis is the default option to store non-critical Red Hat Quay configuration.
  - Quay, the application service itself.

With all prerequisites met, we can install the Quay Operator. See next how to install it.

Operator installation

The process to deploy the Quay Operator is simple. Follow the instructions in the next steps:

1. Navigate to **OperatorHub** and search for Red Hat Quay.

Figure 13.1 – Seach for the Red Hat Quay Operator

2.   Click on the tile and then on the **Install** button.

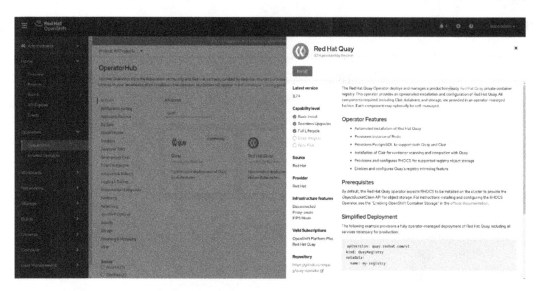

Figure 13.2 – Install Red Hat Quay

3.   Leave the default options and click on the **Install** button.

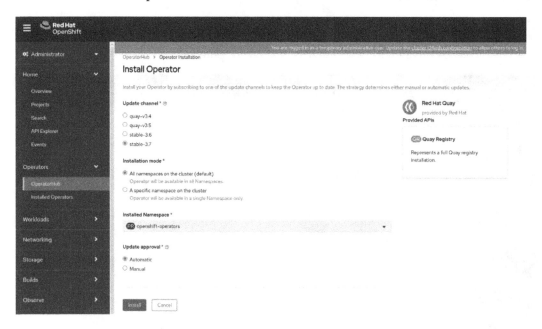

Figure 13.3 – Operator installation

4.   When the Operator is installed, access the **Quay Registry** tab and click on **Create instance**.

Figure 13.4 – Creating a Quay Registry instance

5.   Optionally, change the name in the proper field and leave all default options.

Figure 13.5 – Quay instance creation

6. Check the **Quay Registry** tab. You will see the **Status** column similar to the following screenshot.

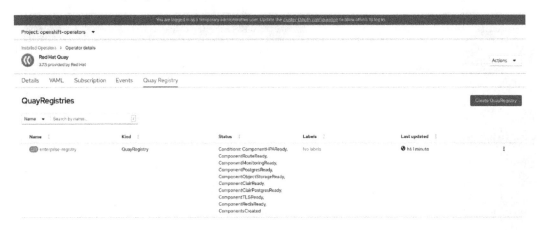

Figure 13.6 – Quay instance created

7. As soon as it finishes, click on the registry that has been just created and inspect the **Registry Endpoint** URL.

Figure 13.7 – Quay endpoint

Click on the registry endpoint address. The Quay Registry application will be shown to configure your multi-cluster registry.

Configuring Quay

Once the installation process finishes, you will have access to some important endpoints that are used to configure the integration between the Quay Operator, the Quay Registry instance, and OpenShift. To use them, you need to follow the next steps:

1. In the namespace in which you installed Quay (in our case, the `openshift-operators` namespace), navigate to **Workloads | Secrets** and search for a secret starting with `<objectInstanceName>-quay-config-editor-credentials-<randomPodId>`.

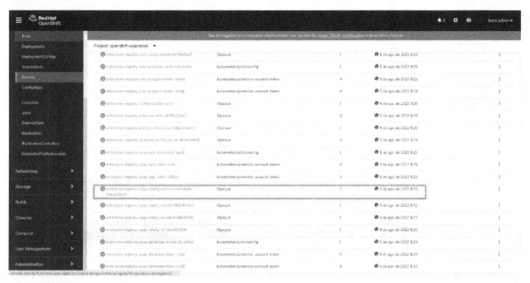

Figure 13.8 – Check Quay config credentials

2. Click on it and choose **Reveal Values**.

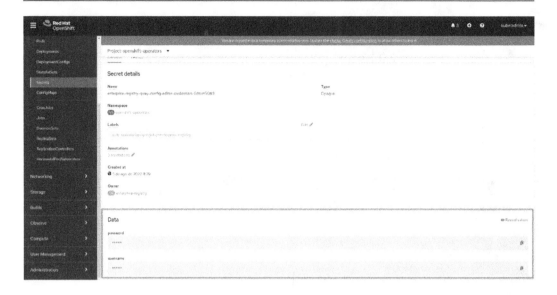

Figure 13.9 – Quay credential values

3. Take note of the credentials and navigate back to **Config Editor Endpoint** in the Quay Registry Operator instance.

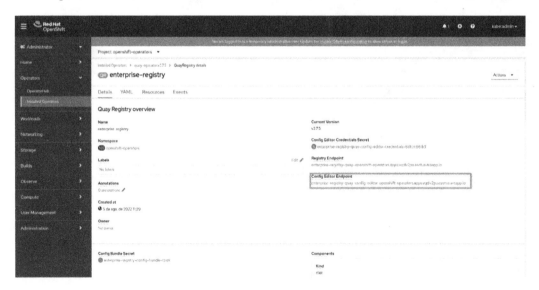

Figure 13.10 – Quay config endpoint

4. Click the link to access **Config Editor Tool**. Use the credentials you got in the previous step.

Figure 13.11 – Quay UI config tool

5. On this page, you can change many Quay configurations, such as SSL certificates, and so on – in our example, we changed the **Time Machine** expiration time. Click on **Validate Configuration Changes** after you make your changes.

Figure 13.12 – Quay settings config tool

6. After the configurations are validated, choose **Reconfigure Quay**.

Validating configuration

✓ Configuration Validated   Continue Editing   Download   Reconfigure Quay

Figure 13.13 – Quay config tool validation

> **Important Note**
>
> This step will restart the operator pods and Quay may become unavailable for a short period.

7. The Quay Config Editor tool will automatically update the `quay-config-bundle` secret, used to store the configuration parameters. To see it navigate to **Workloads | Secrets**, search for the `quay-config-bundle` secret and click on **Reveal values**. The changes made previously must be reflected inside this secret.

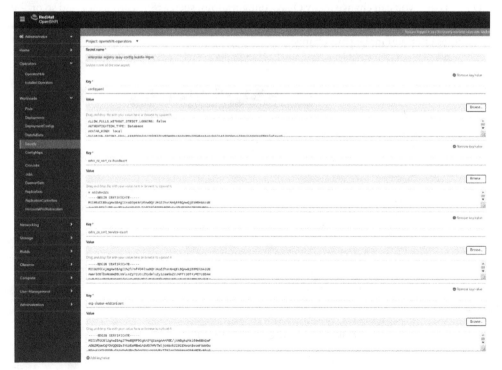

Figure 13.14 – Quay secret after config tool changes

If everything appears correct, you will be able to start using your Quay instance. Proceed with the next section to see instructions on how to use Quay.

Using Red Hat Quay

Red Hat Quay usage is very simple. A user-friendly interface will help you to configure additional global and security settings. We recommend you create an *organization* to organize your Quay registry by departments, regions, or any other division you want:

1. In this example, we created an organization named **multiclusterbook**.

Figure 13.15 – Quay organization

2. Next, we created a private repository to upload our images.

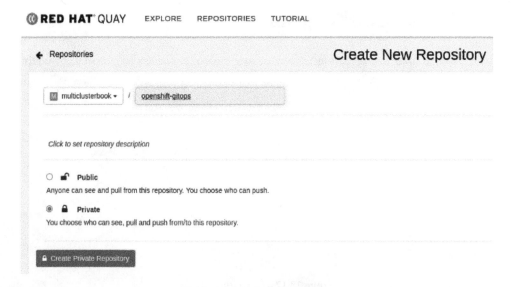

Figure 13.16 – Quay repository

A very helpful way to learn how to use Red Hat Quay is using the tutorial. You'll see how to do that.

Running the tutorial

Running the tutorial mentioned in the Quay UI is not required but it can be a good starting point to understanding Red Hat Quay as an enterprise container registry. The first step of this tutorial is to log in via the terminal to the Quay registry endpoint. We are using **podman** as the container engine but you can alternatively use **Docker** instead if you prefer, both will work similar, so you just need to replace podman with Docker in this case (for example, `docker login` instead of `podman login`, `docker run` instead of `podman run`, and so on):

```
$ podman login enterprise-registry-quay-openshift-operators.
apps.vqdlv2pu.eastus.aroapp.io
```

You should have a successful login, otherwise, check the credentials used. After, try to run a **BusyBox** container image.

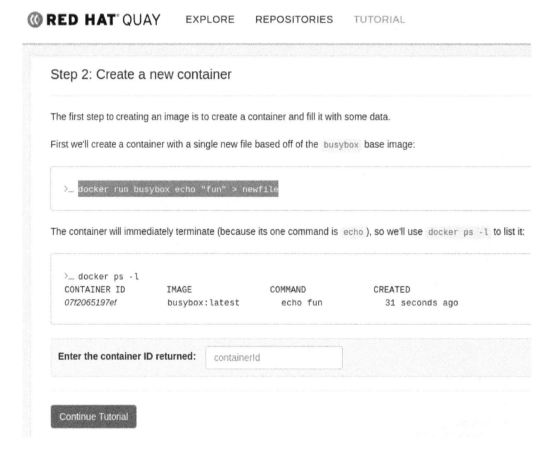

Figure 13.17 – Quay tutorial

Now, run the following commands to create a sample container image for this repository:

```
$ podman run busybox echo "fun" > newfile
$ podman ps -1
CONTAINER ID   IMAGE                                COM-
MAND      CREATED          STA-
TUS                       PORTS         NAMES
fd3b51f4c383   docker.io/library/busybox:latest   echo fun     16
seconds ago   Exited (0) 17 seconds ago                        zealous_
pascal
$ podman commit fd3b51f4c383 enterprise-registry-quay-open-
shift-operators.apps.vqdlv2pu.eastus.aroapp.io/multicluster-
book/openshift-gitops
$ podman push enterprise-registry-quay-openshift-operators.
apps.vqdlv2pu.eastus.aroapp.io/multiclusterbook/openshift-gi-
tops
```

Simple, right?! Now your instance of Quay Container Registry is configured, tested, and ready to use. Start exploring it in your CI/CD pipelines, pushing images to it, and using it as the enterprise container registry for your DevOps and DevSecOps pipelines.

Now that we've already covered Quay, let's discuss a bit more about OpenShift Plus. Next, you'll continue to discover why you should care about and consider OpenShift Plus in your hybrid or multi-cloud strategy.

What is OpenShift Plus?

OpenShift Plus is an offering from Red Hat that bundles the OpenShift platform with the products you can see in the following diagram:

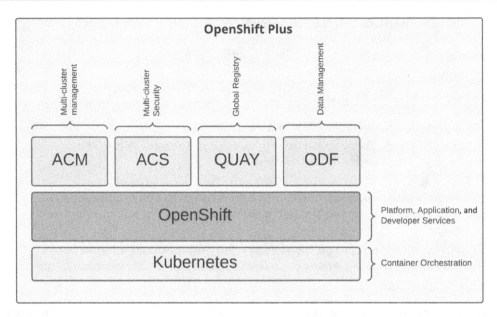

Figure 13.18 – OpenShift Plus products offering

These products are included with this offering(besides OpenShift itself):

- **Advanced Cluster Management**: Provides features to manage several OpenShift and Kubernetes clusters. The features include cluster management, observability, policy and compliance, and workload deployment. We dove into this product in *Chapter 11, OpenShift Multi-Cluster GitOps and Management*. If you haven't seen that chapter yet, we highly encourage you to go back and walk through it now.

- **Advanced Cluster Security**: Enhance the security of your Kubernetes and OpenShift clusters with features such as Vulnerability Management, Network management, and others. This product was covered in *Chapter 12, OpenShift Multi-Cluster Security*. We also encourage you to check this chapter now if you haven't done so yet.

- **Quay Enterprise Registry**: The global enterprise registry was covered at the beginning of this chapter.

- **OpenShift Data Foundation**: This is a storage layer that can provide different types of volumes to OpenShift clusters: file storage (for RWX persistent volumes) using **Cephfs**, block storage (for RWO persistent volumes) using **Ceph RBD**, and object storage provided by a solution named **Noobaa**. This product is not covered in this book. If you want to see more information about it, check the links we left in the *Further reading* section of this chapter.

If you have followed this book from the beginning, this is not new information for you. We will focus now on giving additional information that can help you to understand the value proposition behind OpenShift Plus and why you should consider OpenShift Plus, instead of *just* the OpenShift platform.

Value proposition: benefits

OpenShift Plus adds some great benefits to your multi-cluster strategy:

- **Consistency**: ACM provides the governance feature, which helps you to have consistent configurations among all your clusters. In terms of security, ACS also implements security policies you can define that will also help to keep all clusters compliant with them.

- **Portability**: Quay provides a global image registry that all clusters pull images from. ACM also can help you to deploy applications against many clusters at once and to inspect and monitor their status. By using it together with OpenShift GitOps, you can leverage **ArgoCD** to implement **GitOps** deployment practices and maintain the desired state of your applications, no matter in which OpenShift cluster and cloud provider you are running them. Of course, the portability between clusters also depends on the application, as the more decoupled the application is, the easier it will be to port between different OpenShift clusters and providers. The important thing to understand is that ACM and Quay help to make your applications portable much easier. OpenShift Data Foundation also helps to make stateful workloads easier to port, by providing a standard way for applications to have access to Persistent Volumes, regardless of the infrastructure or cloud providers the clusters are running on.

- **Management**: Managing several clusters in different environments is not an easy task in general. Companies usually have many people focused only on doing that. You can turn this activity into a bit of an easier task and perhaps decrease the costs associated with it by using ACM. Features such as cluster management, observability, and governance (policies) are helpful to manage several clusters from a single pane.

- **Security**: As we discussed in *Chapter 12, OpenShift Multi-Cluster Security*, making a few clusters secure and compliant in general is not a hard task, however, when we scale to several clusters it becomes more challenging. ACS adds the capability to easily enforce security policies, detect vulnerabilities, and plan actions to remediate them.

Besides OpenShift Plus, Red Hat also provides some additional value-added products available in the Red Hat Hybrid Cloud portal (`cloud.redhat.com`) for any OpenShift customer:

- **OpenShift Cluster Management**: This allows you to view high-level cluster information, create new clusters using the Assisted Installer, configure Red Hat subscriptions, basic level monitoring, and more.

- **Cost Management**: Cost management is an interesting tool to track costs and usage of AWS, Google Cloud, Azure cloud providers, and OpenShift clusters as well. It helps you to have a consolidated view of the costs associated with multiple cloud accounts and also across your hybrid cloud infrastructure, to inspect cost trends, and to even use it as a basis for charging departments or teams by their usage. See the link for more information about this tool in the *Further reading* section of this chapter.

In the next screenshot, you see an example of what the **Cost Management Overview** screen looks like.

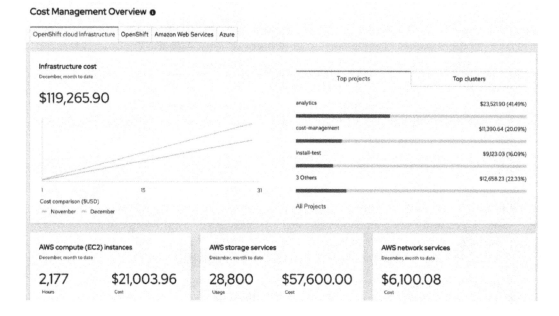

Figure 13.19 – Cost Management feature

- **Developer Sandbox**: This is a free OpenShift environment in a shared, multi-tenant OpenShift cluster pre-configured with a set of developer tools. Use it as a sandbox for your tests and as an environment to learn how to develop for OpenShift. You will find the link for it in the *Further reading* section of this chapter.

Besides the items listed in the preceding list, OpenShift Plus may make sense from the subscription cost as well. Any package is usually priced better than buying individual pieces separately. That being said, if you are interested in any software that is part of the *Plus* stack mentioned in this chapter, it might make sense to buy OpenShift Plus, instead of OpenShift and the software only. Consult your Red Hat account team and ask for a quote for OpenShift Plus.

Now that we've already discussed some of the benefits of having the OpenShift Plus stack and considerations for when it is really compelling, let's use a fictional use case as an example to make it more tangible.

OpenShift Plus – a practical use case

Let's consider the following use case. Try to use the concepts we covered in this book to identify which tool you could use to cover each requirement.

Use case: The company *HybridMyCloud* is developing an AI/ML application that has the following requirements:

- The application life cycle is comprised of Development, QA, Pre-prod, and Production environments.

- Development and QA can share the same underlying platform. Pre-prod and Production need to be separated due to regulatory needs. Production application needs to be in a private subnet, with no public direct access from the internet.

- The C-level and procurement department of *HybridMyCloud* signed a contract with AWS and, due to that, wants to host most resources there. However, they don't want to be locked into it and want to have the ability to move the application to Azure when needed.

- This AI/ML application is based on containers and stateless, however, it stores images in S3 object storage. It receives an X-Ray image, stores it in S3 object storage, and uses an ML-trained model to make a risk assessment of pneumonia.

- This application handles **Protected Health Information** (PHI), therefore *HybridMyCloud* wants to be compliant with the **Health Insurance Portability and Accountability Act** (**HIPAA**).

- This is a mission-critical application, so the enterprise architects of *HybridMyCloud* decided to have this application deployed not only on multi-availability zones but also using multiple regions on AWS.

- Other policies defined by *HybridMyCloud* are as follows:

  - All containers need to have resource requests and limits for CPU and memory.

  - Never use the latest tag images.

  - All images used need to come from the enterprise registry. Pulling images from any other sources other than the enterprise registry and Red Hat certified image registries is strictly prohibited.

So, how would you fulfill all these requirements? There are many different options to match these requirements; OpenShift Plus is definitely a good option that can comply with all of them. Let's discuss what the infrastructure would look like and how OpenShift Plus tools would be used with this use case:

Several OpenShift clusters are required to fulfill these requirements:

- One cluster for Development and QA

- One cluster for Pre-prod

- Two clusters for Production, each one in a different AWS region

- One spare cluster on Azure that can host the application at any given time

You could use the Cluster Lifecycle Management feature of **Advanced Cluster Management** first to deploy these clusters on AWS and Azure. For S3 storage, either the native cloud provider services can be used or, alternatively, use the **OpenShift Data Foundation** with *Multi-Cloud Object Gateway*, which would add the following benefits:

- Use Object Bucket Claims to create S3 volumes.

- Use the *Multi-Cloud Object Gateway* to mirror the volume between AWS and Azure.

With clusters in place, **Advanced Cluster Management** can also be used for the following:

- **Advanced Cluster Management** in conjunction with **OpenShift GitOps** can be used to deploy the application in multiple clusters at the same time.

- Policies can be created to set **LimitRange** objects in all namespaces of each cluster and also edit the `allowedRegistries` field of the `image.config.openshift.io/cluster` custom resource to only allow images that come from the enterprise registry to run.

- Use the observability feature to monitor the application and cluster health.

Quay can be used as the enterprise registry and the embedded image vulnerability scanning tool can be used as one of the sources to find known vulnerabilities. ACS can be used to detect known vulnerabilities in all clusters and assess them using HIPAA policies. It also can be used in the CI/CD pipeline to perform a static security analysis of manifests and to find anti-patterns, such as the usage of the latest tag image.

These are just a few examples. Other tools we covered in this book can also be used, such as OpenShift Pipelines to implement CI/CD pipelines, Cost Management to track costs with cloud providers, and Red Hat Container Catalog to provide the best container base image, and so on.

Summary

In this chapter, we had a brief overview of Red Hat Quay, an interesting option for a container image registry. We discussed the main benefits of OpenShift Plus and why it makes sense to at least consider OpenShift Plus with your hybrid or multi-cloud strategy. Finally, we used a fictional use case to think about how the OpenShift Plus stack could be used in a practical scenario.

Wow, what a great journey so far! We have covered many things already in this book and we are so pleased that you have come along with us on this journey. We are almost at the end of our long journey here, but the best part is coming: in the next chapter, we will exercise most of what we have learned in this book using one practical example. We will have a comprehensive review of OpenShift Pipelines, GitOps, and all other tools we covered in this book using an example, practice with the concepts and tools from scratch, and have a real functional application deployed into multiple clusters in the end.

We encourage you to move on to the next chapter and try the exercises we propose there.

Further reading

Looking for more information? Check the following references to get more information about Red Hat Quay:

- *Quay product documentation:* `https://access.redhat.com/documentation/en-us/red_hat_quay/3.7`

- *Quay Object Storage prerequisite:* `https://access.redhat.com/documentation/en-us/red_hat_quay/3.7/html/deploy_red_hat_quay_on_openshift_with_the_quay_operator/operator-preconfigure#operator-storage-preconfig`

- *Clair – Vulnerability Static Analysis for Containers:* `https://github.com/quay/clair`

Check the following references to get more information about Red Hat OpenShift Data Foundation:

- *OpenShift Data Foundation Smart Page:* `https://www.redhat.com/en/technologies/cloud-computing/openshift-data-foundation`

- *OpenShift Data Foundation Product Page:* `https://access.redhat.com/documentation/en-us/red_hat_openshift_data_foundation`

Check the following references to get more information about Red Hat OpenShift Plus:

- *OpenShift Plus Smart Page:* `https://www.redhat.com/en/technologies/cloud-computing/openshift/platform-plus`

Check the following references to get more information about Red Hat Cost Management:

- *Official documentation:* `https://access.redhat.com/documentation/en-us/cost_management_service/2022`

- *Upstream project:* `https://project-koku.github.io/`

Check the following references to get more information about OpenShift Developer Sandbox:

- *Getting started:* `https://developers.redhat.com/developer-sandbox`

## 14

Building a Cloud-Native Use Case on a Hybrid Cloud Environment

It has been a wonderful journey so far! We walked through so much content in this book already, from OpenShift architecture to Pipelines, GitOps, and multi-cloud tools! We are now reaching our main goal with this book, which is helping you to make the best decisions and implement a good hybrid/multi-cloud strategy for your OpenShift footprint. To wrap up this book with helpful content, we will make a comprehensive review using a practical approach to building and deploying an application using most features we covered during this book: OpenShift Pipelines (Tekton), OpenShift GitOps (ArgoCD), Advanced Cluster Management, Quay, and Advanced Cluster Security.

Therefore, you will find the following in this chapter:

- Use case description
- Application build using OpenShift Pipelines and S2I
- Application deployment using OpenShift Pipelines and GitOps
- Adding security checks in the building and deployment process
- Provisioning and managing multiple clusters
- Deploying an application into multiple clusters

So, what are we waiting for? Let's play now!

> **Note**
> The source code used in this chapter is available at `https://github.com/PacktPublishing/OpenShift-Multi-Cluster-Management-Handbook/tree/main/chapter14`.

Use case description

To be a bit closer to what you see in the real world, this time we are going to use a Java application, using **Quarkus**, which is a great option to build modern, cloud-native applications with Java. Look at the references in the *Further reading* section of this chapter for more information about **Quarkus**.

Our application source code was extracted from the *Getting started with Quarkus* sample; see reference for it in the *Further reading* section of this chapter. During this chapter, we will create a CI/CD pipeline that will do the following:

1.   Build the application using s2i to generate Java binaries.

2.   Push the container image to Quay.

3.   Run a security scan on the image using Advanced Cluster Security.

4.   Deploy the application on the local cluster using ArgoCD.

5.   Deploy the application on multiple remote clusters using ArgoCD and Advanced Cluster Management.

We are going to use Advanced Cluster Management to make all OpenShift clusters compliant with a standard policy we defined for them as well. For the sake of learning and simplicity, we are going to build the pipeline and other objects in sequential phases, like building blocks that are added to build a house.

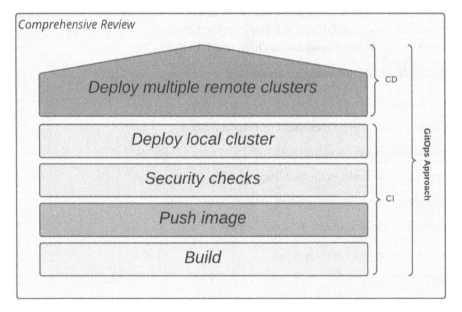

Figure 14.1 – Comprehensive review - Building blocks

We assume that you have access to an OpenShift cluster, which we will call the Hub cluster, with enough resources and with the following tools already installed:

- OpenShift Pipelines

- OpenShift GitOps

- Advanced Cluster Management

- Advanced Cluster Security

- Quay

- The `oc` command line installed and connected to the Hub cluster

We will also deploy additional single node clusters on AWS to be used as managed remote clusters, to exercise the application deployment into multiple clusters. If you haven't installed these tools yet, refer to the installation process of each from *Chapters 9* to *13* of this book.

The source code used in this chapter is available at our GitHub repository: `https://github. com/PacktPublishing/OpenShift-Multi-Cluster-Management-Handbook/ tree/main/chapter14`.

Let's start by digging into the first building block: the application build.

Application build using OpenShift Pipelines and S2I

For this step, we are going to use the `quarkus-build` pipeline that you can find in the `chapter14/ Build/Pipeline/quarkus-build.yaml` file. This pipeline is very straightforward and explained in the following diagram:

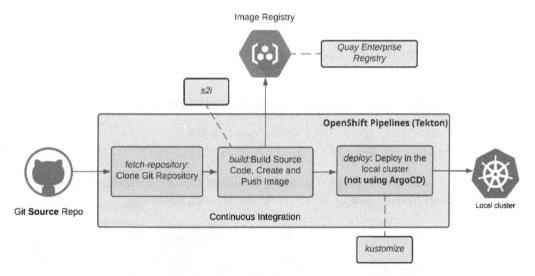

Figure 14.2 – Pipeline to build a Java Quarkus application

In this pipeline we are using pre-existing ClusterTasks to do all the work:

- `git-clone`: Used to clone the Git repository.

- `s2i-java`: Build the Java source code using S2I and Buildah to generate the image and push it to the Quay registry. S2I is a very convenient way to build code from many different languages, such as Java, Python, Node.js, and others. See the *Further reading* section of this chapter for more information about S2I.

- `openshift-client`: Used to run the manifests that deploy the application. Application manifests use **Kustomize** to declare the Kubernetes manifest. We covered **Kustomize** in *Chapter 10, OpenShift GitOps – ArgoCD*, of this book; if you didn't read it yet, we strongly recommend you to do so now and then get back here to perform the steps in this chapter.

Now let's create and run this pipeline. If you haven't done it yet, fork this repository to your GitHub account: `https://github.com/PacktPublishing/OpenShift-Multi-Cluster-Management-Handbook`. After you forked it, follow the instructions in this section to create and run this pipeline:

1. Clone the repository in your machine:

   ```
   $ GITHUB_USER=<your_user>
   $ git clone https://github.com/PacktPublishing/OpenShift-
   Multi-Cluster-Management-Handbook.git
   ```

2. Run the following script and follow the instructions to change the references from the original repository (`PacktPublishing`) to your forked repository:

   ```
   $ cd OpenShift-Multi-Cluster-Management-Handbook/
   chapter14
   $ ./change-repo-urls.sh
   # Go to the Build folder
   $ cd Build
   ```

3. Run the following commands to create the namespace and the pipeline:

   ```
   $ oc apply -f Pipeline/namespace.yaml
   $ oc apply -f Pipeline/quarkus-build-pi.yaml
   ```

4. You should be able to see the pipeline in the OpenShift console, in **Pipelines | Pipelines | Project: chap14-review-cicd**, as you can see in the following screenshot:

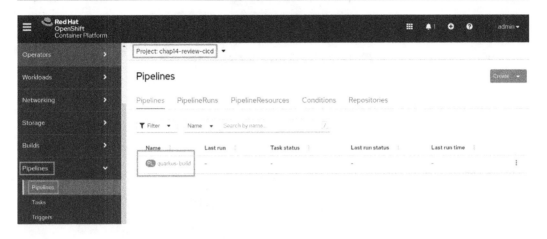

Figure 14.3 – Build pipeline created

5. You can now run the pipeline either using the web interface or through the terminal. To do so from your terminal, run the following command:

```
$ oc create -f PipelineRun/quarkus-build-pr.yaml
```

> **Note about Image Registry**
>
> This pipeline uses an external registry to push the resulting image. To be able to push an image to the registry, you need to link a secret that contains the registry credentials with the `pipeline` ServiceAccount. If you don't do it before running the pipeline, you will notice that it will fail in the `build` task. We are using Quay in this chapter, but you can use any external image registry, such as Nexus, Amazon Elastic Container Registry, Docker Hub, or any other. If you decide to use Quay, you need to create a robot account, give it write permissions in the image repository, and import the secret to the namespace. Next, you will find out how to do it.

See next how to configure your Quay repository and link the credentials to the `pipeline` ServiceAccount.

Configuring the image registry

After you have created a new repository on Quay, follow these steps to configure it:

1. Access the **Settings** tab of the repository and access the **Create robot account** link in the **User and Robot Permissions** section:

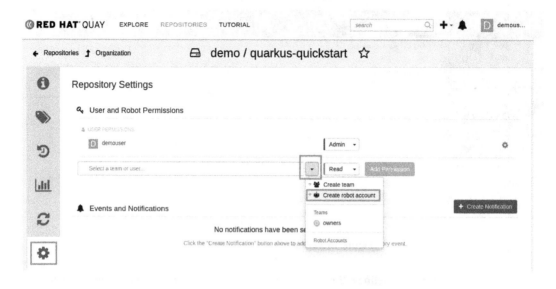

Figure 14.4 – Create a robot account

2.   Give it any name and click on the **Create robot account** button:

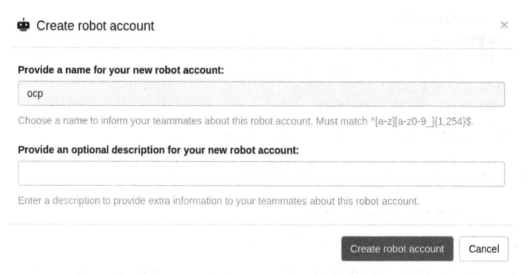

Figure 14.5 – Create a robot account

3.   Then, change the permission to **Write** and click on **Add Permission**:

Figure 14.6 – Set Write permissions

4. Click on the robot account link to download the secret that we will use to link with the pipeline ServiceAccount:

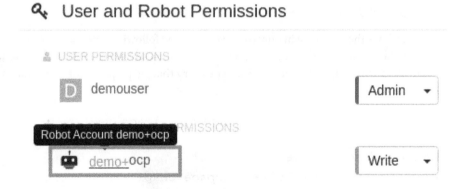

Figure 14.7 – Robot account

5. Download the secret by clicking on the **Download <robot-name>-secret.yml** link in the **Kubernetes Secret** tab:

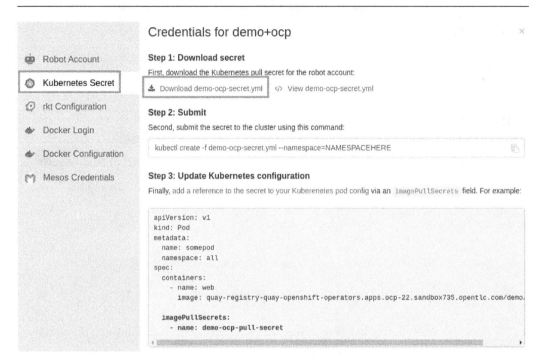

Figure 14.8 – Download Quay credentials

With the secret YAML file in hand, you can proceed with its creation on OpenShift. See next how to do it.

Linking image registry credentials

Now that we already have the secret file in our workspace, run the following commands to create the secret and link it to the pipeline ServiceAccount. Alternatively, you can just run the link-image-registry-secret.sh script from the GitHub repository that we prepared for you, which will do this same process:

```
$ SECRET_FILE=<secret-file-which-contains-image-registry-
credentials>
$ SECRET_NAME=<secret-name>
$ oc create -f $SECRET_FILE --namespace=chap14-review-cicd
$ oc patch serviceaccount pipeline -p '{"secrets": [{"name":
"$SECRET_NAME"}]}' --namespace=chap14-review-cicd
$ oc secrets link default $SECRET_NAME --for=pull -n chap14-
review-cicd
```

If you faced the error mentioned in the build task of the pipeline, you can now run it again by running the following command:

```
$ oc create -f PipelineRun/quarkus-build-pr.yaml
```

Now you should see the pipeline finishing successfully, as you can see in the following screenshot:

Figure 14.9 – Build pipeline run successfully

After the pipeline runs successfully, you may want to see what the image looks like on Quay.

Checking the image on Quay

If you are using Quay, at this stage, you should be able to see and inspect the image there:

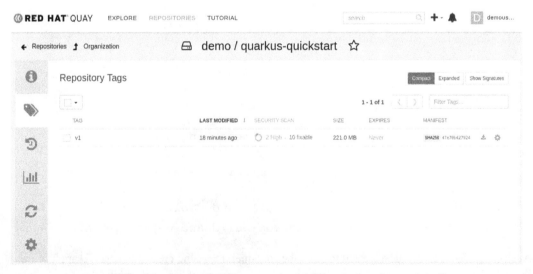

Figure 14.10 – Image on Quay, known vulnerabilities detected automatically

As we can see, Quay detected automatically that this image has some known vulnerabilities. We are going to fix these vulnerabilities further in this chapter, but it is important now that you observe and understand how easy it is to push images and start checking them automatically against known vulnerabilities.

Now that we already have the building pipeline of our application working, let's evolve it to use ArgoCD as the deployment tool, leveraging GitOps practices.

Application deployment using OpenShift Pipelines and GitOps

This time, we are going to use ArgoCD to deploy the application instead of directly running the Kubernetes manifests. The pipeline is basically the same, but now the deploy task will run a YAML file that creates an ArgoCD application and wait until the application becomes healthy.

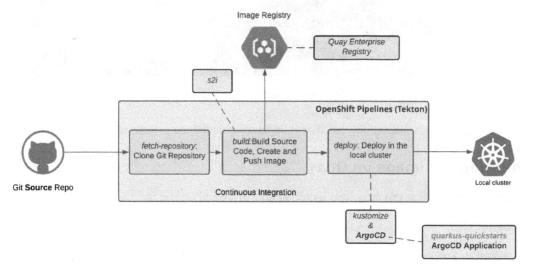

Figure 14.11 – Pipeline to build a Java Quarkus application and deploy it using ArgoCD

Run the following command to create and run the pipeline:

```
$ cd OpenShift-Multi-Cluster-Management-Handbook/chapter14/
Deploy
$ oc apply -f Rolebindings/ # Permission required for Tekton to
create an ArgoCD application
$ oc apply -f Pipeline/quarkus-build-and-deploy-pi.yaml
$ oc create -f PipelineRun/quarkus-build-and-deploy-pr.yaml
```

A new `PipelineRun` will be created to build the container image and create the ArgoCD application that will deploy the application. You will see the following if everything works well:

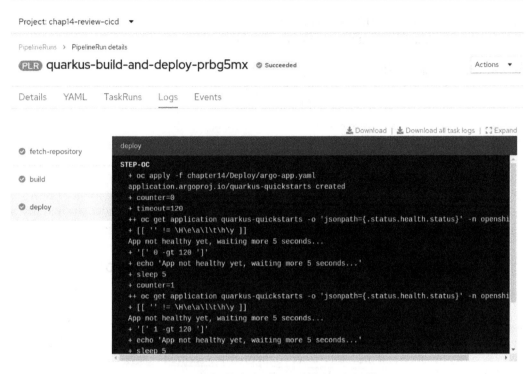

Figure 14.12 – Task deployment using ArgoCD

Access the ArgoCD console to check the application deployment from there; you should see something similar to the following screenshot:

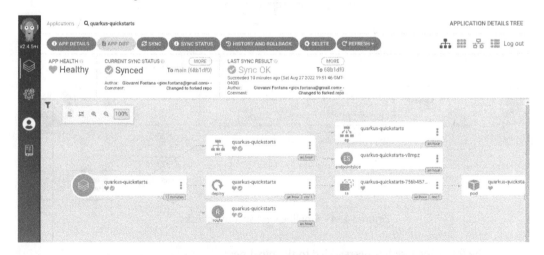

Figure 14.13 – Application in ArgoCD

You can find instructions about how to access the ArgoCD console in *Chapter 10, OpenShift GitOps – ArgoCD*. As a reminder, see next the commands to get the ArgoCD URL and admin password:

```
# Get the ArgoCD URL:
$ echo "$(oc get route openshift-gitops-server -n openshift-
gitops --template='https://{{.spec.host}}')"
# Get the Admin password
$ oc extract secret/openshift-gitops-cluster -n openshift-
gitops --to=-
```

Now our pipeline already builds the application, pushes it to Quay, and deploys it using ArgoCD. The next step is to bring Advanced Cluster Security to add a security check step in our pipeline. See next how to do it.

Adding security checks in the building and deployment process

This time, we will add a new step to perform a security check in the image that has been built. We are going to use Advanced Cluster Security for that. To successfully use it, you should have Advanced Cluster Security installed and the local cluster configured as a secured cluster. Check *Chapter 12, OpenShift Multi-Cluster Security*, to see how to do it.

See next what our pipeline looks like now:

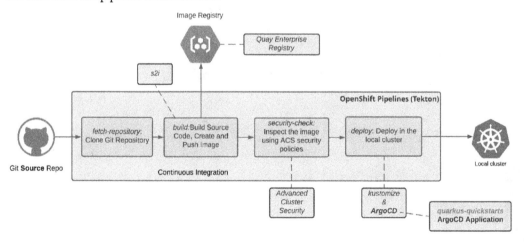

Figure 14.14 – Pipeline with security checks

Therefore, the following task has been added to the pipeline:

- `security-check`: Uses ACS APIs to check the image against existing security policies defined in ACS.

To simulate security issues, we will also use a custom `s2i-java` task that uses an old `ubi-openjdk` version, which contains many known vulnerabilities. To fix the issues, we will change the build strategy to use a Dockerfile that uses the latest version of the RHEL UBI image and additional security fixes.

Follow the instructions in this section to create and run this pipeline:

1. Before we get into the pipeline, we need to configure the integration between the pipeline and ACS. To do so, access the **Advanced Cluster Security** dashboard and navigate to **Platform Configuration | Integrations | Authentication Tokens**, and click on **API Token**:

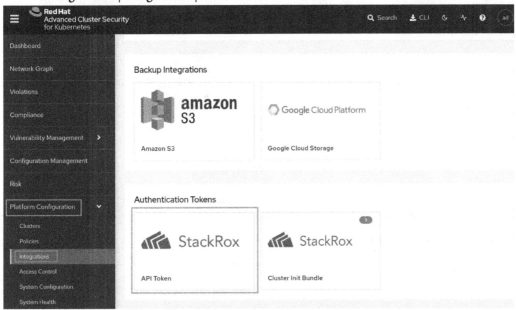

Figure 14.15 – Creating ACS API Token

2. Click on the **Generate token** button:

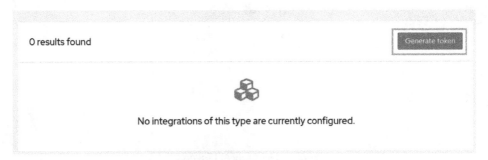

Figure 14.16 – Generate token

3. Fill out a name and select **Continuous Integration** in the **Role** field:

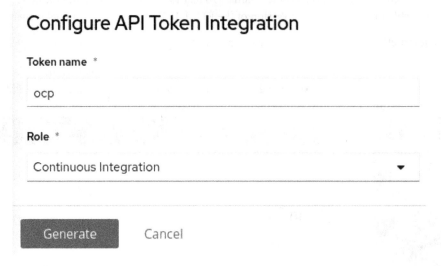

Figure 14.17 – Generate token for CI

4. Copy the token that has been generated. We are going to use it in a secret that will be used by the pipeline task to authenticate on ACS APIs:

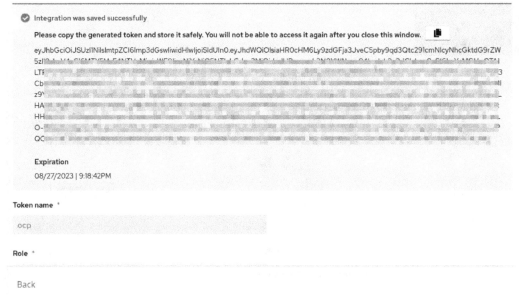

Figure 14.18 – Copy API Token

5. Now let's create the secret. Run the following command using the token from the previous step and the ACS central endpoint. Do *not* use `http(s)` in the `rox_central_endpoint` host:

```
oc create secret generic acs-secret --from-literal=rox_
api_token='<token from previous step>' --from-
literal=rox_central_endpoint='<host-for-rhacs-central-
server>:443' --namespace=chap14-review-cicd
```

6. Now we are all set to create and run our pipeline. Run the following commands:

```
$ cd OpenShift-Multi-Cluster-Management-Handbook/
chapter14/DevSecOps
$ oc apply -f Task/ # Create the custom S2I and stackrox-
image-check tasks
$ oc apply -f Pipeline/quarkus-devsecops-v1-pi.yaml
$ oc create -f PipelineRun/quarkus-devsecops-v1-pr.yaml
```

7. You should see failures in the `security-check` task as we are intentionally using an old base image that contains many known vulnerabilities:

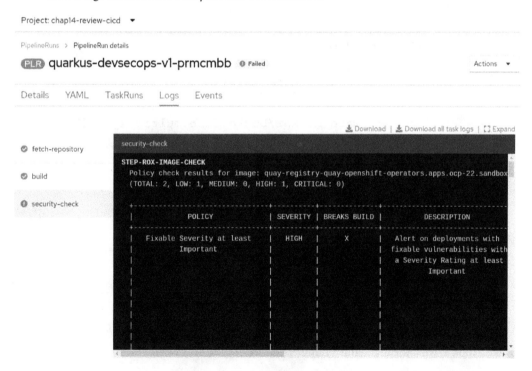

Figure 14.19 – Security checks failure

Let's take a look briefly at the errors we have as a result of this task. The policies that have failed are the following:

- **Fixable Severity at least Important**: As expected (remember that we are using now an old base image version), there are several components in the image that have important and critical known CVEs.

- **Red Hat Package Manager in Image**: Our S2I image uses `ubi-minimal`, which includes `microdnf` as a package manager.

We are going to demonstrate now how to fix these security issues using a Dockerfile that addresses all of them.

Fixing security issues

To fix the issues, we are going to change our pipeline to use a Dockerfile instead of the S2I. To do so we changed the `build` task to use the `buildah` ClusterTask instead of `s2i-java`:

```
- name: build
  params:
    - name: IMAGE
      value: $(params.IMAGE_NAME)
    - name: CONTEXT
      value: $(params.PATH_CONTEXT)
    - name: DOCKERFILE
      value: src/main/docker/Dockerfile.multistage #[1]
    - name: TLSVERIFY
      value: 'false'
  runAfter:
    - fetch-repository
  taskRef:
    kind: ClusterTask
    name: buildah #[2]
  workspaces:
    - name: source
      workspace: workspace
```

Let's take a look at what the highlighted numbers mean:

- **[1]**: The path where the Dockerfile with security fixes is located

- **[2]**: `buildah` ClusterTasks that will build the application using the given Dockerfile

Let's also take a look at the Dockerfile to understand the security fixes. This file is located at `quarkus-getting-started/src/main/docker/Dockerfile.multistage` in our GitHub:

```
## Stage 1: build with maven builder image with native
capabilities
FROM quay.io/quarkus/ubi-quarkus-native-image:22.2-java17 AS
build
(.. omitted ..)
RUN ./mvnw package -Pnative
FROM registry.access.redhat.com/ubi8/ubi-minimal #[1]
WORKDIR /work/
COPY --from=build /code/target/*-runner /work/application
RUN chmod 775 /work /work/application \
   && chown -R 1001 /work \
   && chmod -R "g+rwX" /work \
   && chown -R 1001:root /work \
   && microdnf update -y \ #[2]
   && rpm -e --nodeps $(rpm -qa '*rpm*' '*dnf*' '*libsolv*'
'*hawkey*' 'yum*') #[3]
EXPOSE 8080
USER 1001
CMD ["./application", "-Dquarkus.http.host=0.0.0.0"]
```

Let's take a look at what the highlighted numbers mean:

- **[1]**: Use the latest version of `ubi-minimal` as the base image.
- **[2]**: Update OS packages to the latest versions.
- **[3]**: Remove the package manager from the image.

The lines highlighted will make sure the most up-to-date components, which contain the most recent security fixes, are in use, and also the package manager is removed from the image before it is packaged.

Now, let's create this new pipeline version and runs it to check whether the security issues have been resolved. To do so, run the following commands:

```
$ oc apply -f Pipeline/quarkus-devsecops-v2-pi.yaml
$ oc create -f PipelineRun/quarkus-devsecops-v2-pr.yaml
```

This time, the pipeline should be finished successfully, as there are no security issues detected anymore in our container image:

Figure 14.20 – Security issues fixed

You can optionally check ACS now to investigate whether there are still other violations that may be fixed later. If you want to do so, navigate to the **Violations** feature of ACS and filter by Namespace: chap14-review-cicd and Deployment: quarkus-quickstarts. You should still see some minor violations, as follows:

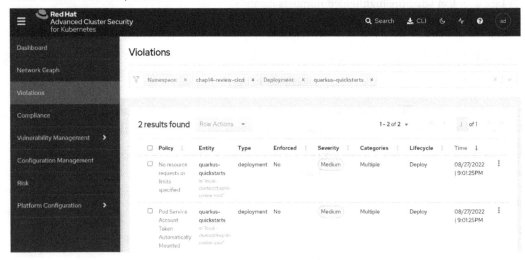

Figure 14.21 – ACS Violations

Do you remember that Quay reported some vulnerabilities in our image before? Look at it now to see our new image version:

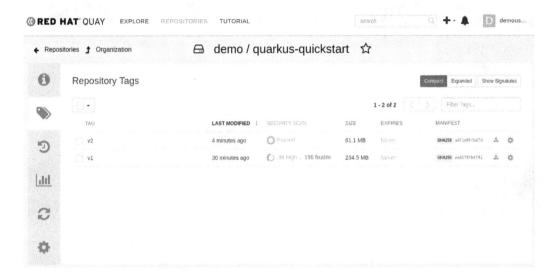

Figure 14.22 – Quay security scan

As you can see, the newer image has no security issues detected. In this section, we added a security check in our pipeline and fixed some vulnerabilities detected by this pipeline. In the next section, our pipeline will be able to deploy our application against multiple clusters, using ArgoCD and Advanced Cluster Management.

Provisioning and managing multiple clusters

We haven't touched so far on the hybrid or multi-cluster side of the house. This is what we are going to add now: *deployment into multiple remote clusters*. To do so, we are going to use Advanced Cluster Management to provision new clusters and also help us to deploy the application in them.

Provisioning new clusters

We are going to use AWS to host two new clusters that will be used as remote clusters to exercise our pipeline. For the sake of saving resources, we are going to use single node clusters, so we don't need to get the cost of many servers for this exercise. If you already have clusters available, you can alternatively import the existing clusters, instead of provisioning new ones. You can find, in the *Further reading* section of this chapter, a link that contains instructions about how to import a cluster on Advanced Cluster Management.

To provision a single node cluster using ACM, you need to add the AWS credentials, navigate to the **Credentials** menu, and click on the **Add credential** button:

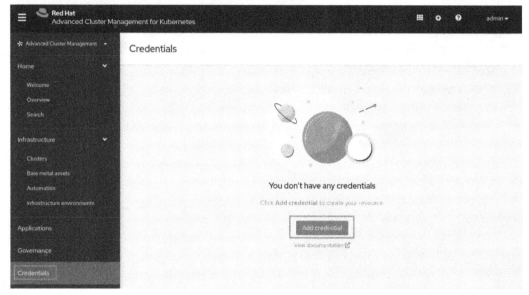

Figure 14.23 – Adding AWS credentials

Follow the wizard and fill out all required fields. You need to provide your pull secret, which is available at `https://console.redhat.com/openshift/downloads`:

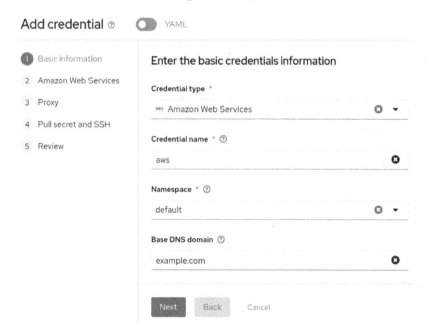

Figure 14.24 – Adding a new credential

After you have created the AWS credential, access the **Infrastructure | Clusters** feature and click on the **Create cluster** button:

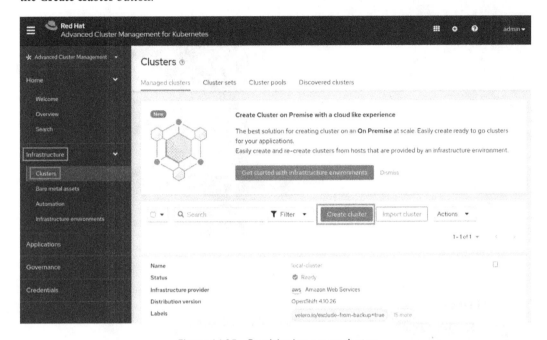

Figure 14.25 – Provisioning a new cluster

Select AWS as the infrastructure provider and fill out the wizard with the following data but do *not* hit the **Create** button in the last step of the wizard:

- **Infrastructure** provider: **Amazon Web Services**
- **Infrastructure provider credential: aws** (name of the credential that you created in the previous step)
- **Cluster name**: `ocp-prd1`
- **Cluster set: default**
- **Base DNS domain**: Your public domain on AWS (for example, example.com)
- **Release image**: Select the most recent
- **Additional labels**: `env=prod`
- **Control plane pool**: Instance type: `m5.2xlarge`
- **Networking**: Leave as-is
- **Proxy**: Leave unselected
- **Automation**: Leave blank

On the **Review** page, select the **YAML: On** button:

Figure 14.26 – Edit YAML

In the YAML file, edit `MachinePool` and add the statement `skipMachinePool: true`, as you can see in the following:

Figure 14.27 – Editing MachinePool

Click in the `install-config` tab and change master replicas to 1 and compute replicas to 0:

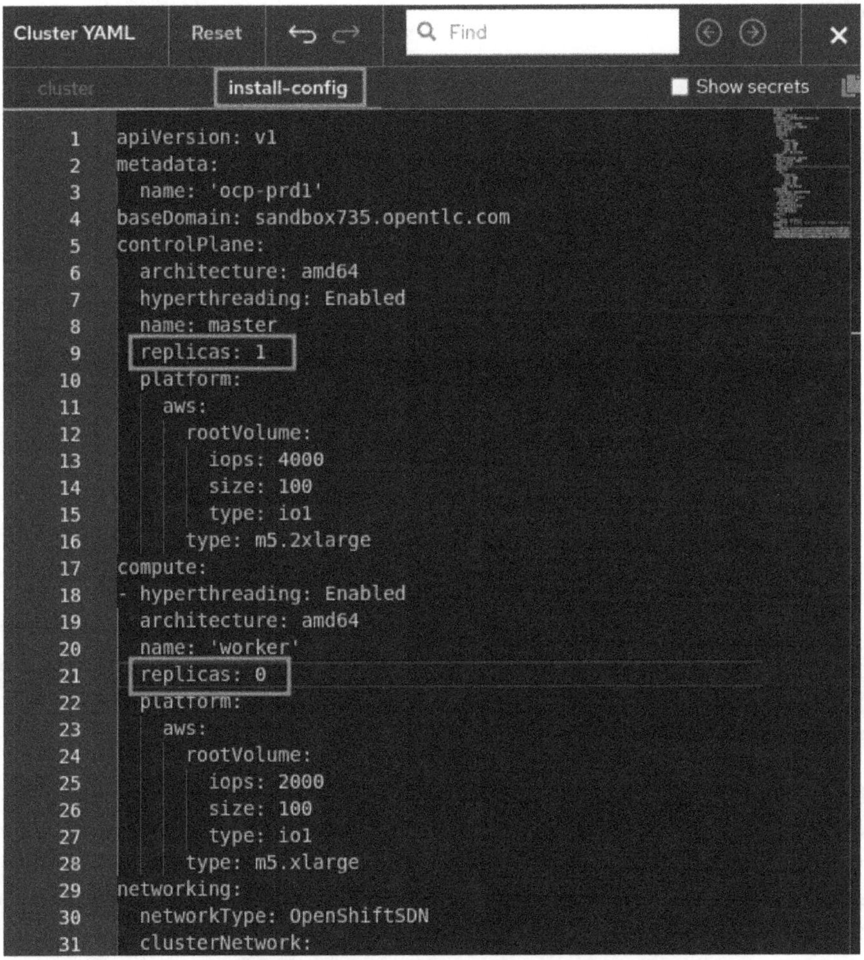

Figure 14.28 – Editing install-config

Now hit the **Create** button. Repeat the steps to create another cluster named `ocp-prd2` with the same parameters used previously. In the end, you should see two clusters being provisioned, as you can see in the following screenshot:

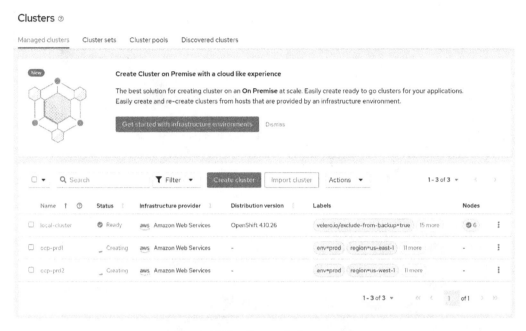

Figure 14.29 – Clusters being created

The provisioning will take about 40 minutes. Continue next when you see both clusters marked as **Ready**.

Cluster governance

One helpful feature that ACM provides is cluster governance using policies. We already covered this feature in *Chapter 11*, *OpenShift Multi-Cluster GitOps and Management*. If you didn't read it yet, we strongly recommend you check that chapter. We are going to deploy the policy that is in the `Governance` folder of our GitHub repository to inform if the etcd keystores of managed clusters are encrypted or not. To do so, run the following command:

```
$ cd OpenShift-Multi-Cluster-Management-Handbook/chapter14/
Governance
$ oc apply -k .
```

Wait a few seconds and access the **Governance** feature on ACM to check the compliance of the clusters:

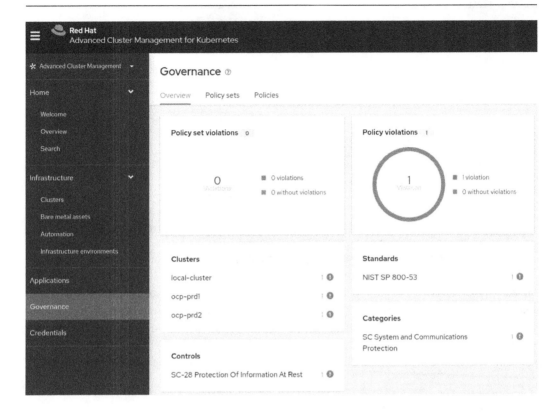

Figure 14.30 – Cluster compliance

Move to the next section to see how to deploy our sample application into multiple remote clusters at once.

Deploying an application into multiple clusters

Now that we already have multiple remote clusters, we can go ahead and use ACM and ArgoCD to make our pipeline able to deploy into all of them at once. We are going to change the deploy task to use an `ApplicationSet` object that will be responsible for deploying our application into both OpenShift remote clusters at once.

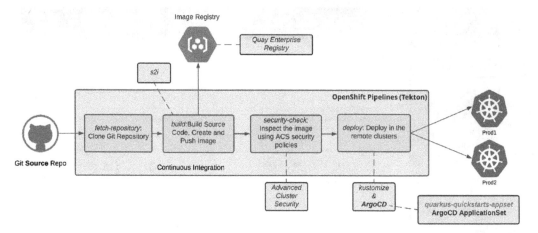

Figure 14.31 – Pipeline with deployment into multiple clusters

To make ArgoCD aware of the clusters managed by ACM, we first need to create a few objects, such as the `GitOpsCluster` Custom Resource. We covered a detailed explanation of these objects in *Chapter 11, OpenShift Multi-Cluster GitOps and Management*. Run the following commands to create these objects:

```
$ cd OpenShift-Multi-Cluster-Management-Handbook/chapter14/
Multicluster-Deployment
$ oc apply -f GitOpsCluster/
```

Now let's create and run the pipeline, which uses an `ApplicationSet` object to deploy the application into the managed clusters that have the `env=prod` label. Remember that we used this label in the clusters we provisioned using ACM. If you imported the clusters on ACM, make sure to add the `env=prod` label to them:

```
$ oc apply -f Rolebindings/ # Permissions required for pipeline
to be able to create an ApplicationSet
$ oc apply -f Pipeline/quarkus-multicluster-pi.yaml
$ oc create -f PipelineRun/quarkus-multicluster-pr.yaml
```

When the pipeline finishes, you should now have two new ArgoCD applications automatically created by the `ApplicationSet` mechanism:

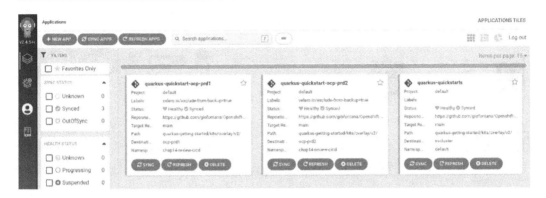

Figure 14.32 – ArgoCD and ApplicationSet

That's it, we did it! We started with a simple build pipeline that now performs security checks and deploys into multiple remote clusters at once!

Summary

Thank you for being our partner in this journey! We hope the content of this book was helpful for you and now you have a good understanding of the topics covered in this book. We went through architecture, people, deployment, troubleshooting, multi-cluster administration, usage, and security. So much content that we thought we wouldn't have the ability to write it! And if you are still here, we feel that my mission with this book is accomplished!

There is a quote from *Johann Wolfgang von Goethe* that says the following: "*Knowing is not enough; we must apply. Willing is not enough; we must do.*" After reading this book, we hope you not only learned new things but were also able to put them into practice. Following this hybrid cloud journey, you have the opportunity to leap in knowledge with didactic examples and content made with great dedication from us.

We hope that this book becomes one of your handbooks and will be useful to you for planning and executing models suitable for the enterprise, bringing multiple options to use, implementations, and good insights to leverage your knowledge and your career.

To wrap up the content of this chapter, we designed the following diagram to serve as a shortcut to the central themes of each chapter and see the entire journey we have gone through together:

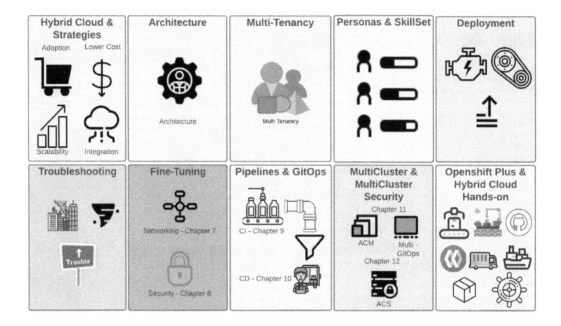

Figure 14.33 – The book journey

We have almost reached the end of this book, but we are not completely done yet. We have prepared for you the last chapter with some suggestions as to where you can go next after this book, to keep learning and growing your OpenShift and Kubernetes skills.

We encourage you to move to the next chapter and look at the training and other content we suggest there.

Further reading

Looking for more information? Check the following references to get more information:

- *Quarkus*:

    - *Main page:* https://quarkus.io/

    - *Quarkus getting started sample:* https://github.com/quarkusio/quarkus-quickstarts/tree/main/getting-started

- *S2i:*

    - *GitHub* page: https://github.com/openshift/source-to-image

    - *How to Create an S2I Builder Image* (blog article): https://cloud.redhat.com/blog/create-s2i-builder-image

- *Using Source 2 Image build in Tekton* (blog article): `https://cloud.redhat.com/blog/guide-to-openshift-pipelines-part-2-using-source-2-image-build-in-tekton`

- *Tekton Hub - S2I*: `https://hub.tekton.dev/tekton/task/s2i`

- *Advanced Cluster Management – Importing clusters*: `https://access.redhat.com/documentation/en-us/red_hat_advanced_cluster_management_for_kubernetes/2.5/html/clusters/managing-your-clusters#importing-a-target-managed-cluster-to-the-hub-cluster`

- *Advanced Cluster Security – Image check from Tekton Hub*: `https://hub.tekton.dev/tekton/task/stackrox-image-check`

Part 5 – Continuous Learning

In this part, you will have some additional content to continue the journey to becoming a subject matter expert on OpenShift. Here, you will see a summary of the training available on the market to help you with OpenShift's in-depth enablement.

This part of the book comprises the following chapter:

- *Chapter 15, What's Next?*

15
What's Next

Learning never exhausts the mind.

– Leonardo da Vinci

The last chapter! When we were writing this book, we were wondering what a good last chapter for it would be. We strongly believe that learning never ends! That being said, we came to the conclusion that a good last chapter would be related to some suggestions about training and the next steps you could take to keep learning and deepening your knowledge even further—we hope you enjoy the options we introduce here.

In particular, we'll look at the following areas:

- Red Hat training
- Online learning platforms
- Free training and references

Red Hat training

It is evident the benefits that any organization gets from training and enablement. In a nutshell, training is capable of increasing productivity, reducing waste, standardizing processes, and boosting inner motivation. **International Data Corporation (IDC)** research from 2020 in five different countries and companies with more than 11,000 employees on average found the following benefits from taking training provided by Red Hat:

- **Development-operations (DevOps)** teams experienced 44% higher productivity.
- 34% more efficient **information technology (IT)** infrastructure teams compared to those that had not taken the training.
- Teams deploy new IT resources 59% faster.
- New employees achieve full productivity 76% faster.

However, there is a large set of different training and resources available on the market, for *every pocket*! For this reason, we brought here some suggestions about where to look for enablement and educate you and your team on OpenShift. Combine the references you find in this chapter with the personas you saw in *Chapter 4, OpenShift Personas and Skillset*, and you will likely have a successful and great OpenShift adoption!

You will find the following in this chapter:

- A list of suggested training resources

- References on the internet where you can look for documentation, tutorials, and posts about OpenShift

- A list of certifications that will help you and your team to establish as experts with OpenShift

> **Reference**
> Check out the complete IDC research mentioned previously at `https://red.ht/IDCTraining`.

Training and certifications

This book is intended to be a comprehensive reference for OpenShift, and we covered some important topics around architecture, deployment, operations, and multi-cluster tools. However, we can confidently say that OpenShift is a world of different cultures! As such, although this book will help you in your journey, we strongly recommend that you also educate yourself through training. In this section, you will find some skill paths we suggest for professionals that want to go even deeper with their OpenShift knowledge.

Skills paths

We bring to you here three different suggested skills paths:

- **DevOps transformation**: For DevOps and **site reliability engineering (SRE)** engineers that are looking into working with DevOps and Agile practices and tools on top of OpenShift. As we have seen in *Chapter 4, OpenShift Personas and Skillset*, this role usually works with automation and processes, so in this skills path, you will find courses that address people, processes, and technology related to DevOps.

- **Kubernetes**: This is a path for those who want to have more in-depth information about Kubernetes itself. As OpenShift runs on top of Kubernetes, the training covered in this path is also valuable for OpenShift; however, it is agnostic and not focused on any distribution of Kubernetes, such as OpenShift and others.

- **OpenShift**: Training that covers both sides of OpenShift usage, development, and administration. In this section, you will see a list of training and certifications suggested for each profile.

Check out the following sections for more information on each skill path.

DevOps transformation

In this section, we suggest training that will take you through aspects related to people, processes, and technology.

DevOps Culture and Practice Enablement

We know that DevOps has been adopted for quite some time and almost everywhere; however, from our personal experience, there are several companies that unfortunately don't realize the benefits of its adoption. This is mostly due to the cultural change that these companies often forget to implement: cultural shift is the most challenging thing about DevOps—by far, from our perspective. If you feel that you and your team need a cultural refreshment, this course is for you. It follows a practical approach with 5-day immersive training that we recommend you take with your teammates to experience near-real-world delivery using agile methodologies and DevOps practices. Here's a course overview:

Course name	DevOps Culture and Practice Enablement—TL500
Vendor	Red Hat (paid)
Duration	5 days (classroom or on-site)
Pre-requisites	Familiarity with agile is helpful, but not required.
More information	`https://www.redhat.com/en/services/training/do500-devops-culture-and-practice-enablement`

Open Practices for your DevOps Journey

While the previous training is focused on culture, this one is focused on processes. It takes students through the discovery, planning, and delivery of projects driven by DevOps practices. Here's an overview of the course:

Course name	Open Practices for your DevOps Journey—TL250
Vendor	Red Hat (paid)
Duration	4 days (classroom, virtual, or on-site)
Pre-requisites	None
More information	`https://www.redhat.com/en/services/training/red-hat-training-open-practices-your-devops-journey-do250`

Red Hat DevOps Pipelines and Processes: CI/CD with Jenkins, Git, and Test Driven Development

To close the loop of people (culture), processes, and technology, this training will build technical skills for DevOps teams around **continuous integration/continuous delivery (CI/CD)** technologies using Git, Jenkins, and **test-driven development (TDD)**. It is a practical training course that uses sample applications to learn how to use the technologies listed. Have a look at an overview of the course here:

Course name	Red Hat DevOps Pipelines and Processes: CI/CD with Jenkins, Git, and Test Driven Development—DO400
Vendor	Red Hat (paid)
Duration	4 days (classroom, virtual, on-site, or on-demand)
Pre-requisites	Familiarity with software development is required.
More information	`https://www.redhat.com/en/services/training/do400-red-hat-devops-pipelines-and-processes-with-jenkins-git-and-test-driven-development`

Kubernetes

You will easily find a huge collection of training courses about Kubernetes on the market, paid and free ones. We outlined here some good references so that you can go deeper into Kubernetes itself.

Kubernetes Basics

The first we want to indicate is a tutorial from the official `kubernetes.io` portal. In this tutorial, you will be guided through some basic lessons such as Kubernetes cluster creating using `minikube`, application deployment, scaling, and updates. It uses interactive environments so that you can exercise the lessons step by step interactively with a practical approach. Here's an overview of the course:

Course name	Kubernetes Basics
Vendor	`kubernetes.io` (free)
Duration	2-4 hours (on-demand)
Pre-requisites	None
More information	`https://kubernetes.io/docs/tutorials/kubernetes-basics/`

Another great practical reference is a project named *Kube by Example*, which we will give more details about next.

Kube by Example

This project contains several practical examples to help you learn more about Kubernetes. You have access to lessons about basic concepts such as what is a pod, deployment, service, namespace, and so on, but also learning paths related to application development and more. Have a look at an overview of the course here:

Course name	Kube by Example
Vendor	Red Hat (free)
Duration	1-4 hours each lesson (on-demand)
Pre-requisites	None
More information	`https://kubebyexample.com/`

Certifications

There are three official Kubernetes certifications, each one focused on validating a different set of skills, as you can see in the following table:

	Certified Kubernetes Administrator (CKA)	Certified Kubernetes Application Developer (CKAD)	Certified Kubernetes Security Specialist (CKS)
Description	Focused on administration tasks, such as installation, configuration, and management of production-grade clusters.	This exam looks to validate the examinees' skills related to the design, build, and deployment of cloud-native applications on Kubernetes.	Exam dedicated to security aspects of the development of applications and management using Kubernetes.
Recommended for	System/cloud/platform administrators; DevOps/SRE engineers	Application developers and DevOps/SRE engineers	Security engineers
Exam duration	2 hours	2 hours	2 hours
More information	`https://training.linuxfoundation.org/certification/certified-kubernetes-administrator-cka/`	`https://training.linuxfoundation.org/certification/certified-kubernetes-application-developer-ckad/`	`https://training.linuxfoundation.org/certification/certified-kubernetes-security-specialist/`

OpenShift

Red Hat has many great training courses and certifications you can take to get in-depth practical knowledge about OpenShift. The following diagram shows training and certifications recommended by Red Hat that will give a comprehensive in-depth knowledge of Linux, containers, and OpenShift administration and development:

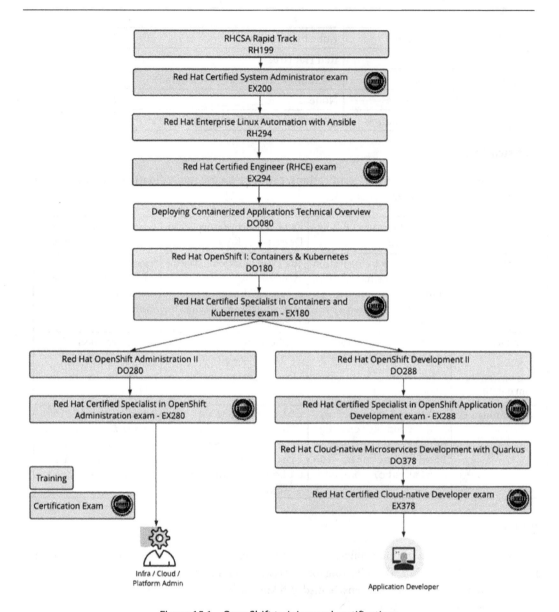

Figure 15.1 – OpenShift training and certifications

Refer to this site for more details about the training courses and certifications: `https://www.redhat.com/en/services/training/all-courses-exams`

Red Hat Learning Subscription (7-day free trial)

If you are interested in taking the mentioned OpenShift skill path, a good option is **Red Hat Learning Subscription (RHLS)**! With that, you can get access to hundreds of online courses with practical labs. Some options include only courses, while others also include certifications. Find more information at this website: `https://www.redhat.com/en/services/training/learning-subscription`.

You may also refer to online learning platforms to learn more; see the next section for some suggestions.

Online learning platforms

We couldn't forget to also mention some of the online learning platforms out there. Today, there is an enormous amount of good resources on online learning platforms; you can find a small list of them here:

- *A Cloud Guru*: `https://acloudguru.com/browse-training`
- *Pluralsight*: `https://www.pluralsight.com/`
- *Cloud Academy*: `https://cloudacademy.com/search/`
- *Whizlabs*: `https://www.whizlabs.com/`

If you are looking for more free training, check out the ideas presented next.

Free training and references

What if my pocket is empty and I can't afford to pay for training? We didn't forget you either! Next, we'll see some of the options to learn OpenShift for free.

OpenShift Container Platform Demo and Workshop Guide

A great free reference is the *OpenShift Container Platform Demo and Workshop Guide* page. Here, you will find some video recordings of demos and also links for lab documentation, and more. You will find a wide variety of demos about cluster installation, **artificial intelligence (AI)/machine learning (ML)**, application development, GitOps, and security. You can find more information here:

Reference	OpenShift Container Platform Demo and Workshop Guide
Vendor	Red Hat (free)
Duration	Miscellaneous (on-demand)
Pre-requisites	Familiarity with OpenShift is helpful.
More information	`https://demo.openshift.com`

OpenShift 4 101 Workshop

The Red Hat Services team made a great job of documenting fantastic workshops about Red Hat products. One of them is the OpenShift 4 101 workshop, which is excellent training for OpenShift and covers application development, deployment and management, CI/CD pipeline, and others. Here's further information on this:

Reference	OpenShift 4 101
Vendor	Red Hat (free)
Duration	Miscellaneous (on-demand)
Pre-requisites	Familiarity with OpenShift is helpful. Access to an OpenShift cluster is required.
More information	`https://redhatgov.io/workshops/openshift_4_101/`

Open Demos

A good source of technical information is webinars and demos. On the *Open Demos* page, you can find a schedule of the next free webinar that anyone can attend. More information is available here:

Reference	Open Demos
Vendor	Red Hat (free)
Duration	Miscellaneous (on-demand)
Pre-requisites	Familiarity with OpenShift is helpful.
More information	`https://redhatdemos.com/`

Red Hat Developer portal

The Red Hat Developer portal also contains a great number of good and free references, from free tutorials to a developer sandbox. Find out more here:

Reference	Red Hat Developer portal
Vendor	Red Hat (free)
Duration	Miscellaneous (on-demand)
Pre-requisites	Familiarity with OpenShift is helpful.
More information	`https://developers.redhat.com/products/openshift/getting-started`

YouTube and Twitch

You will find great content on *YouTube* and *Twitch*. The Red Hat product management team, for instance, runs a *What's New in OpenShift* session each quarter, detailing the main innovations the next version will introduce. This is a great source of information about what is coming next in the product. Find out more here:

Reference	YouTube and Twitch
Vendor	Red Hat (free)
Duration	Miscellaneous (on-demand)
Pre-requisites	Familiarity with OpenShift is helpful.
More information	`https://www.youtube.com/c/redhat`, `https://mobb.ninja/` `https://www.twitch.tv/redhatopenshift`

Blogs

Blogs are great references to be up to date and also find some great demos and labs. The following are great blogs about OpenShift:

* `https://cloud.redhat.com/blog`
* `https://developers.redhat.com`

Product documentation

Always refer to the OpenShift documentation to learn about its features and clear any doubts. The official documentation can be found at the following website: `https://docs.openshift.com`.

Summary

We have seen in this chapter a list of suggested training, skill paths, and certifications to get a deeper knowledge of OpenShift. We also brought to you some free references, if you don't want to (or can't!) pay for training. If you are interested in establishing yourself as a **subject-matter expert** (**SME**) in it, we encourage you to consider them.

Final words

Congratulations—you reached the end of this book! We made it! We really hope you now feel more prepared and confident about working with OpenShift in the cloud, on-premises, and even both—from one cluster to many! Thanks for being our partner during the course of this book, and we look forward to meeting you anytime soon!

Our sincere message of thanks and, as we said, learning is constant and the sources of information are many. To achieve your goals without getting lost or straying along the path of so many opportunities, we leave some tips here:

- Look at the technology options available today.

- Research those that have seen the most adoption and absorption in terms of ease of use, learning, and documentation, as well as employment and career opportunities.

- Choose the way forward and stay focused.

- Emerging technologies can be a good path, but they are often risky. If you want to adopt new technologies, don't forget to take into account the path and plan you created, as this will help you to continue understanding where you are and where you want to go.

- Write down your findings—they will surely help you very soon.

- During your studies, set aside leisure time with family and friends.

- Exercise regularly—a healthy mind also comes from a healthy body.

  "Mens sana in corpore sano." – Decimus Iunius Iuvenalis

As the Kubernetes and OpenShift ecosystem continues to evolve each day, check the OpenShift website and blog for the latest features and keep yourself informed about updates and roadmaps. If the company where you work is assisted by a Red Hat account team, reach out to them frequently. Red Hat has a great team to assist customers, and they are continuously looking to make their customers comfortable and satisfied with their products, so don't hesitate to use them as your ally.

Now, it is time to say goodbye—or better: see you out there! If you have questions or ideas for improvement, or just want to say hello, reach out to us on our social networks, GitHub, or the *Packt Community* platform. Thank you again and we hope you enjoyed the *OpenShift Multi-Cluster Management Handbook*!

Index

Packt.com

Subscribe to our online digital library for full access to over 7,000 books and videos, as well as industry leading tools to help you plan your personal development and advance your career. For more information, please visit our website.

Why subscribe?

- Spend less time learning and more time coding with practical eBooks and Videos from over 4,000 industry professionals

- Improve your learning with Skill Plans built especially for you

- Get a free eBook or video every month

- Fully searchable for easy access to vital information

- Copy and paste, print, and bookmark content

Did you know that Packt offers eBook versions of every book published, with PDF and ePub files available? You can upgrade to the eBook version at packt.com and as a print book customer, you are entitled to a discount on the eBook copy. Get in touch with us at customercare@packtpub.com for more details.

At www.packt.com, you can also read a collection of free technical articles, sign up for a range of free newsletters, and receive exclusive discounts and offers on Packt books and eBooks.

Other Books You May Enjoy

If you enjoyed this book, you may be interested in these other books by Packt:

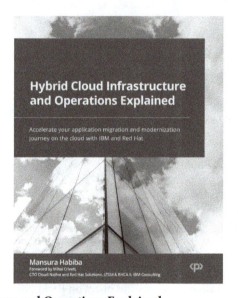

Hybrid Cloud Infrastructure and Operations Explained

Mansura Habiba

ISBN: 9781803248318

- Strategize application modernization, from the planning to the implementation phase
- Apply cloud-native development concepts, methods, and best practices
- Select the right strategy for cloud adoption and modernization
- Explore container platforms, storage, network, security, and operations
- Manage cloud operations using SREs, FinOps, and MLOps principles
- Design a modern data insight hub on the cloud

Building CI/CD Systems Using Tekton

Joel Lord

ISBN: 9781801078214

- Understand the basic principles behind CI/CD
- Explore what tasks are and how they can be made reusable and flexible
- Focus on how to use Tekton objects to compose a robust pipeline
- Share data across a pipeline using volumes and workspaces
- Discover more advanced topics such as WhenExpressions and Secrets to build complex pipelines
- Understand what Tekton Triggers are and how they can be used to automate CI/CD pipelines
- Build a full CI/CD pipeline that automatically deploys an application to a Kubernetes cluster when an update is done to a code repository

Packt is searching for authors like you

If you're interested in becoming an author for Packt, please visit authors.packtpub.com and apply today. We have worked with thousands of developers and tech professionals, just like you, to help them share their insight with the global tech community. You can make a general application, apply for a specific hot topic that we are recruiting an author for, or submit your own idea.

Share Your Thoughts

Now you've finished *OpenShift Multi-Cluster Management Handbook*, we'd love to hear your thoughts! Scan the QR code below to go straight to the Amazon review page for this book and share your feedback or leave a review on the site that you purchased it from.

https://packt.link/r/1803235284

Your review is important to us and the tech community and will help us make sure we're delivering excellent quality content.

Download a free PDF copy of this book

Thanks for purchasing this book!

Do you like to read on the go but are unable to carry your print books everywhere?

Is your eBook purchase not compatible with the device of your choice?

Don't worry, now with every Packt book you get a DRM-free PDF version of that book at no cost.

Read anywhere, any place, on any device. Search, copy, and paste code from your favorite technical books directly into your application.

The perks don't stop there, you can get exclusive access to discounts, newsletters, and great free content in your inbox daily

Follow these simple steps to get the benefits:

1. Scan the QR code or visit the link below

https://packt.link/free-ebook/9781803235288

2. Submit your proof of purchase
3. That's it! We'll send your free PDF and other benefits to your email directly

www.ingramcontent.com/pod-product-compliance
Lightning Source LLC
Chambersburg PA
CBHW081458050326
40690CB00015B/2840